The Next Wave of AI Self-Driving Cars

Practical Innovations in
Artificial Intelligence and Machine Learning

Dr. Lance B. Eliot, MBA, PhD

ISBN: 0578415100
ISBN-13: 978-0578415109

DEDICATION

To my incredible daughter, Lauren, and my incredible son, Michael.

Forest fortuna adiuvat (from the Latin; good fortune favors the brave).

CONTENTS

Lance B. Eliot

ACKNOWLEDGMENTS

I have been the beneficiary of advice and counsel by many friends, colleagues, family, investors, and many others. I want to thank everyone that has aided me throughout my career. I write from the heart and the head, having experienced first-hand what it means to have others around you that support you during the good times and the tough times.

To Warren Bennis, one of my doctoral advisors and ultimately a colleague, I offer my deepest thanks and appreciation, especially for his calm and insightful wisdom and support.

To Mark Stevens and his generous efforts toward funding and supporting the USC Stevens Center for Innovation.

To Lloyd Greif and the USC Lloyd Greif Center for Entrepreneurial Studies for their ongoing encouragement of founders and entrepreneurs.

To Peter Drucker, William Wang, Aaron Levie, Peter Kim, Jon Kraft, Cindy Crawford, Jenny Ming, Steve Milligan, Chis Underwood, Frank Gehry, Buzz Aldrin, Steve Forbes, Bill Thompson, Dave Dillon, Alan Fuerstman, Larry Ellison, Jim Sinegal, John Sperling, Mark Stevenson, Anand Nallathambi, Thomas Barrack, Jr., and many other innovators and leaders that I have met and gained mightily from doing so.

Thanks to Ed Trainor, Kevin Anderson, James Hickey, Wendell Jones, Ken Harris, DuWayne Peterson, Mike Brown, Jim Thornton, Abhi Beniwal, Al Biland, John Nomura, Eliot Weinman, John Desmond, and many others for their unwavering support during my career.

And most of all thanks as always to Michael and Lauren, for their ongoing support and for having seen me writing and heard much of this material during the many months involved in writing it. To their patience and willingness to listen.

Lance B. Eliot

INTRODUCTION

This is a book that provides the newest innovations and the latest Artificial Intelligence (AI) advances about the emerging nature of AI-based autonomous self-driving driverless cars. Via recent advances in Artificial Intelligence (AI) and Machine Learning (ML), we are nearing the day when vehicles can control themselves and will not require and nor rely upon human intervention to perform their driving tasks (or, that <u>allow</u> for human intervention, but only *require* human intervention in very limited ways).

Similar to my other related books, which I describe in a moment and list the chapters in the Appendix A of this book, I am particularly focused on those advances that pertain to self-driving cars. The phrase "autonomous vehicles" is often used to refer to any kind of vehicle, whether it is ground-based or in the air or sea, and whether it is a cargo hauling trailer truck or a conventional passenger car. Though the aspects described in this book are certainly applicable to all kinds of autonomous vehicles, I am focused more so here on cars.

Indeed, I am especially known for my role in aiding the advancement of self-driving cars, serving currently as the Executive Director of the Cybernetic Self-Driving Cars Institute.. In addition to writing software, designing and developing systems and software for self-driving cars, I also speak and write quite a bit about the topic. This book is a collection of some of my more advanced essays. For those of you that might have seen my essays posted elsewhere, I have updated them and integrated them into this book as one handy cohesive package.

You might be interested in companion books that I have written that provide additional key innovations and fundamentals about self-driving cars. Those books are entitled **"Introduction to Driverless Self-Driving Cars,"** **"Advances in AI and Autonomous Vehicles: Cybernetic Self-Driving Cars,"** **"Self-Driving Cars: "The Mother of All AI Projects,"** **"Innovation and Thought Leadership on Self-Driving Driverless Cars,"** **"New Advances in AI Autonomous Driverless Self-Driving Cars,"** and **"Autonomous Vehicle Driverless Self-Driving Cars and**

1

Artificial Intelligence," "Transformative Artificial Intelligence Driverless Self-Driving Cars," "Disruptive Artificial Intelligence and Driverless Self-Driving Cars, and **"State-of-the-Art AI Driverless Self-Driving Cars,"** and **"Top Trends in AI Self-Driving Cars,"** and **"AI Innovations and Self-Driving Cars," "Crucial Advances for AI Driverless Cars," "Sociotechnical Insights and AI Driverless Cars," "Pioneering Advances for AI Driverless Cars"** and **"Leading Edge Trends for AI Driverless Cars,"** and **"The Cutting Edge of Autonomous Cars"** (they are all available via Amazon). See Appendix A of this herein book to see a listing of the chapters covered in those three books.

For the introduction here to this book, I am going to borrow my introduction from those companion books, since it does a good job of laying out the landscape of self-driving cars and my overall viewpoints on the topic. The remainder of the book is all new material that does not appear in the companion books.

INTRODUCTION TO SELF-DRIVING CARS

This is a book about self-driving cars. Someday in the future, we'll all have self-driving cars and this book will perhaps seem antiquated, but right now, we are at the forefront of the self-driving car wave. Daily news bombards us with flashes of new announcements by one car maker or another and leaves the impression that within the next few weeks or maybe months that the self-driving car will be here. A casual non-technical reader would assume from these news flashes that in fact we must be on the cusp of a true self-driving car.

Here's a real news flash: We are still quite a distance from having a true self-driving car. It is years to go before we get there.

Why is that? Because a true self-driving car is akin to a moonshot. In the same manner that getting us to the moon was an incredible feat, likewise can it be said for achieving a true self-driving car. Anybody that suggests or even brashly states that the true self-driving car is nearly here should be viewed with great skepticism. Indeed, you'll see that I often tend to use the word "hogwash" or "crock" when I assess much of the decidedly *fake news* about self-driving cars. Those of us on the inside know that what is often reported to the outside is malarkey. Few of the insiders are willing to say so. I have no such hesitation.

Indeed, I've been writing a popular blog post about self-driving cars and hitting hard on those that try to wave their hands and pretend that we are on the imminent verge of true self-driving cars. For many years, I've been known as the AI Insider. Besides writing about AI, I also develop AI software. I do what I describe. It also gives me insights into what others that are doing AI are really doing versus what it is said they are doing.

Many faithful readers had asked me to pull together my insightful short essays and put them into another book, which you are now holding in your hands.

For those of you that have been reading my essays over the years, this collection not only puts them together into one handy package, I also updated the essays and added new material. For those of you that are new to the topic of self-driving cars and AI, I hope you find these essays approachable and informative. I also tend to have a writing style with a bit of a voice, and so you'll see that I am times have a wry sense of humor and also like to poke at conformity.

As a former professor and founder of an AI research lab, I for many years wrote in the formal language of academic writing. I published in referred journals and served as an editor for several AI journals. This writing here is not of the nature, and I have adopted a different and more informal style for these essays. That being said, I also do mention from time-to-time more rigorous material on AI and encourage you all to dig into those deeper and more formal materials if so interested.

I am also an AI practitioner. This means that I write AI software for a living. Currently, I head-up the Cybernetics Self-Driving Car Institute, where we are developing AI software for self-driving cars. I am excited to also report that my son, also a software engineer, heads-up our Cybernetics Self-Driving Car Lab. What I have helped to start, and for which he is an integral part, ultimately he will carry long into the future after I have retired. My daughter, a marketing whiz, also is integral to our efforts as head of our Marketing group. She too will carry forward the legacy now being formulated.

For those of you that are reading this book and have a penchant for writing code, you might consider taking a look at the open source code available for self-driving cars. This is a handy place to start learning how to develop AI for self-driving cars. There are also many new educational courses spring forth.

There is a growing body of those wanting to learn about and develop self-driving cars, and a growing body of colleges, labs, and other avenues by which you can learn about self-driving cars.

This book will provide a foundation of aspects that I think will get you ready for those kinds of more advanced training opportunities. If you've already taken those classes, you'll likely find these essays especially interesting as they offer a perspective that I am betting few other instructors or faculty offered to you. These are challenging essays that ask you to think beyond the conventional about self-driving cars.

THE MOTHER OF ALL AI PROJECTS

In June 2017, Apple CEO Tim Cook came out and finally admitted that Apple has been working on a self-driving car. As you'll see in my essays, Apple was enmeshed in secrecy about their self-driving car efforts. We have only been able to read the tea leaves and guess at what Apple has been up to. The notion of an iCar has been floating for quite a while, and self-driving engineers and researchers have been signing tight-lipped Non-Disclosure Agreements (NDA's) to work on projects at Apple that were as shrouded in mystery as any military invasion plans might be.

Tim Cook said something that many others in the Artificial Intelligence (AI) field have been saying, namely, the creation of a self-driving car has got to be the mother of all AI projects. In other words, it is in fact a tremendous moonshot for AI. If a self-driving car can be crafted and the AI works as we hope, it means that we have made incredible strides with AI and that therefore it opens many other worlds of potential breakthrough accomplishments that AI can solve.

Is this hyperbole? Am I just trying to make AI seem like a miracle worker and so provide self-aggrandizing statements for those of us writing the AI software for self-driving cars? No, it is not hyperbole. Developing a true self-driving car is really, really, really hard to do. Let me take a moment to explain why. As a side note, I realize that the Apple CEO is known for at times uttering hyperbole, and he had previously said for example that the year 2012 was "the mother of all years," and he had said that the release of iOS 10 was "the mother of all releases" – all of which does suggest he likes to use the handy "mother of" expression. But, I assure you, in terms of true self-driving cars, he has hit the nail on the head. For sure.

When you think about a moonshot and how we got to the moon, there are some identifiable characteristics and those same aspects can be applied to creating a true self-driving car. You'll notice that I keep putting the word "true" in front of the self-driving car expression. I do so because as per my essay about the various levels of self-driving cars, there are some self-driving cars that are only somewhat of a self-driving car. The somewhat versions are ones that require a human driver to be ready to intervene. In my view, that's not a true self-driving car. A true self-driving car is one that requires no human driver intervention at all. It is a car that can entirely undertake via automation the driving task without any human driver needed. This is the essence of what is known as a Level 5 self-driving car. We are currently at the Level 2 and Level 3 mark, and not yet at Level 5.

Getting to the moon involved aspects such as having big stretch goals, incremental progress, experimentation, innovation, and so on. Let's review how this applied to the moonshot of the bygone era, and how it applies to the self-driving car moonshot of today.

Big Stretch Goal

Trying to take a human and deliver the human to the moon, and bring them back, safely, was an extremely large stretch goal at the time. No one knew whether it could be done. The technology wasn't available yet. The cost was huge. The determination would need to be fierce. Etc. To reach a Level 5 self-driving car is going to be the same. It is a big stretch goal. We can readily get to the Level 3, and we are able to see the Level 4 just up ahead, but a Level 5 is still an unknown as to if it is doable. It should eventually be doable and in the same way that we thought we'd eventually get to the moon, but when it will occur is a different story.

Incremental Progress

Getting to the moon did not happen overnight in one fell swoop. It took years and years of incremental progress to get there. Likewise for self-driving cars. Google has famously been striving to get to the Level 5, and pretty much been willing to forgo dealing with the intervening levels, but most of the other self-driving car makers are doing the incremental route. Let's get a good Level 2 and a somewhat Level 3 going. Then, let's improve the Level 3 and get a somewhat Level 4 going. Then, let's improve the Level 4 and finally arrive at a Level 5. This seems to be the prevalent way that we are going to achieve the true self-driving car.

Experimentation

You likely know that there were various experiments involved in perfecting the approach and technology to get to the moon. As per making incremental progress, we first tried to see if we could get a rocket to go into space and safety return, then put a monkey in there, then with a human, then we went all the way to the moon but didn't land, and finally we arrived at the mission that actually landed on the moon. Self-driving cars are the same way. We are doing simulations of self-driving cars. We do testing of self-driving cars on private land under controlled situations. We do testing of self-driving cars on public roadways, often having to meet regulatory requirements including for example having an engineer or equivalent in the car to take over the controls if needed. And so on. Experiments big and small are needed to figure out what works and what doesn't.

Innovation

There are already some advances in AI that are allowing us to progress toward self-driving cars. We are going to need even more advances. Innovation in all aspects of technology are going to be required to achieve a true self-driving car. By no means do we already have everything in-hand that we need to get there. Expect new inventions and new approaches, new algorithms, etc.

Setbacks

Most of the pundits are avoiding talking about potential setbacks in the progress toward self-driving cars. Getting to the moon involved many setbacks, some of which you never have heard of and were buried at the time so as to not dampen enthusiasm and funding for getting to the moon. A recurring theme in many of my included essays is that there are going to be setbacks as we try to arrive at a true self-driving car. Take a deep breath and be ready. I just hope the setbacks don't completely stop progress. I am sure that it will cause progress to alter in a manner that we've not yet seen in the self-driving car field. I liken the self-driving car of today to the excitement everyone had for Uber when it first got going. Today, we have a different view of Uber and with each passing day there are more regulations to the ride sharing business and more concerns raised. The darling child only stays a darling until finally that child acts up. It will happen the same with self-driving cars.

SELF-DRIVING CARS CHALLENGES

But what exactly makes things so hard to have a true self-driving car, you might be asking. You have seen cruise control for years and years. You've lately seen cars that can do parallel parking. You've seen YouTube videos of Tesla drivers that put their hands out the window as their car zooms along the highway, and seen to therefore be in a self-driving car. Aren't we just needing to put a few more sensors onto a car and then we'll have in-hand a true self-driving car? Nope.

Consider for a moment the nature of the driving task. We don't just let anyone at any age drive a car. Worldwide, most countries won't license a driver until the age of 18, though many do allow a learner's permit at the age of 15 or 16. Some suggest that a younger age would be physically too small to reach the controls of the car. Though this might be the case, we could easily adjust the controls to allow for younger aged and thus smaller stature. It's not their physical size that matters. It's their cognitive development that matters.

To drive a car, you need to be able to reason about the car, what the car can and cannot do. You need to know how to operate the car. You need to know about how other cars on the road drive. You need to know what is allowed in driving such as speed limits and driving within marked lanes. You need to be able to react to situations and be able to avoid getting into accidents. You need to ascertain when to hit your brakes, when to steer clear of a pedestrian, and how to keep from ramming that motorcyclist that just cut you off.

Many of us had taken courses on driving. We studied about driving and took driver training. We had to take a test and pass it to be able to drive. The point being that though most adults take the driving task for granted, and we often "mindlessly" drive our cars, there is a significant amount of cognitive effort that goes into driving a car. After a while, it becomes second nature. You don't especially think about how you drive, you just do it. But, if you watch a novice driver, say a teenager learning to drive, you suddenly realize that there is a lot more complexity to it than we seem to realize.

Furthermore, driving is a very serious task. I recall when my daughter and son first learned to drive. They are both very conscientious people. They wanted to make sure that whatever they did, they did well, and that they did not harm anyone. Every day, when you get into a car, it is probably around 4,000 pounds of hefty metal and plastics (about two tons), and it is a lethal weapon. Think about it. You drive down the street in an object that weighs two tons and with the engine it can accelerate and ram into anything you want to hit. The damage a car can inflict is very scary. Both my children were surprised that they were being given the right to maneuver this monster of a beast that could cause tremendous harm entirely by merely letting go of the steering wheel for a moment or taking your eyes off the road.

In fact, in the United States alone there are about 30,000 deaths per year by auto accidents, which is around 100 per day. Given that there are about 263 million cars in the United States, I am actually more amazed that the number of fatalities is not a lot higher. During my morning commute, I look at all the thousands of cars on the freeway around me, and I think that if all of them decided to go zombie and drive in a crazy maniac way, there would be many people dead. Somehow, incredibly, each day, most people drive relatively safely. To me, that's a miracle right there. Getting millions and millions of people to be safe and sane when behind the wheel of a two ton mobile object, it's a feat that we as a society should admire with pride.

So, hopefully you are in agreement that the driving task requires a great deal of cognition. You don't' need to be especially smart to drive a car, and we've done quite a bit to make car driving viable for even the average dolt. There isn't an IQ test that you need to take to drive a car. If you can read and write, and pass a test, you pretty much can legally drive a car. There are of course some that drive a car and are not legally permitted to do so, plus there are private areas such as farms where drivers are young, but for public roadways in the United States, you can be generally of average intelligence (or less) and be able to legally drive.

This though makes it seem like the cognitive effort must not be much. If the cognitive effort was truly hard, wouldn't we only have Einstein's that could drive a car? We have made sure to keep the driving task as simple as we can, by making the controls easy and relatively standardized, and by having roads that are relatively standardized, and so on. It is as though Disneyland has put their Autopia into the real-world, by us all as a society agreeing that roads will be a certain way, and we'll all abide by the various rules of driving.

A modest cognitive task by a human is still something that stymies AI. You certainly know that AI has been able to beat chess players and be good at other kinds of games. This type of narrow cognition is not what car driving is about. Car driving is much wider. It requires knowledge about the world, which a chess playing AI system does not need to know. The cognitive aspects of driving are on the one hand seemingly simple, but at the same time require layer upon layer of knowledge about cars, people, roads, rules, and a myriad of other "common sense" aspects. We don't have any AI systems today that have that same kind of breadth and depth of awareness and knowledge.

As revealed in my essays, the self-driving car of today is using trickery to do particular tasks. It is all very narrow in operation. Plus, it currently assumes that a human driver is ready to intervene. It is like a child that we have taught to stack blocks, but we are needed to be right there in case the child stacks them too high and they begin to fall over. AI of today is brittle, it is narrow, and it does not approach the cognitive abilities of humans. This is why the true self-driving car is somewhere out in the future.

Another aspect to the driving task is that it is not solely a mind exercise. You do need to use your senses to drive. You use your eyes a vision sensors to see the road ahead. You vision capability is like a streaming video, which your brain needs to continually analyze as you drive. Where is the road? Is there a pedestrian in the way? Is there another car ahead of you? Your senses are relying a flood of info to your brain. Self-driving cars are trying to do the same, by using cameras, radar, ultrasound, and lasers. This is an attempt at mimicking how humans have senses and sensory apparatus.

Thus, the driving task is mental and physical. You use your senses, you use your arms and legs to manipulate the controls of the car, and you use your brain to assess the sensory info and direct your limbs to act upon the controls of the car. This all happens instantly. If you've ever perhaps gotten something in your eye and only had one eye available to drive with, you suddenly realize how dependent upon vision you are. If you have a broken foot with a cast, you suddenly realize how hard it is to control the brake pedal and the accelerator. If you've taken medication and your brain is maybe sluggish, you suddenly realize how much mental strain is required to drive a car.

An AI system that plays chess only needs to be focused on playing chess. The physical aspects aren't important because usually a human moves the chess pieces or the chessboard is shown on an electronic display. Using AI for a more life-and-death task such as analyzing MRI images of patients, this again does not require physical capabilities and instead is done by examining images of bits.

Driving a car is a true life-and-death task. It is a use of AI that can easily and at any moment produce death. For those colleagues of mine that are developing this AI, as am I, we need to keep in mind the somber aspects of this. We are producing software that will have in its virtual hands the lives of the occupants of the car, and the lives of those in other nearby cars, and the lives of nearby pedestrians, etc. Chess is not usually a life-or-death matter.

Driving is all around us. Cars are everywhere. Most of today's AI applications involve only a small number of people. Or, they are behind the scenes and we as humans have other recourse if the AI messes up. AI that is driving a car at 80 miles per hour on a highway had better not mess up. The consequences are grave. Multiply this by the number of cars, if we could put magically self-driving into every car in the USA, we'd have AI running in the 263 million cars. That's a lot of AI spread around. This is AI on a massive scale that we are not doing today and that offers both promise and potential peril.

There are some that want AI for self-driving cars because they envision a world without any car accidents. They envision a world in which there is no car congestion and all cars cooperate with each other. These are wonderful utopian visions.

They are also very misleading. The adoption of self-driving cars is going to be incremental and not overnight. We cannot economically just junk all existing cars. Nor are we going to be able to affordably retrofit existing cars. It is more likely that self-driving cars will be built into new cars and that over many years of gradual replacement of existing cars that we'll see the mix of self-driving cars become substantial in the real-world.

In these essays, I have tried to offer technological insights without being overly technical in my description, and also blended the business, societal, and economic aspects too. Technologists need to consider the nontechnological impacts of what they do. Non-technologists should be aware of what is being developed.

We all need to work together to collectively be prepared for the enormous disruption and transformative aspects of true self-driving cars. We all need to be involved in this mother of all AI projects.

WHAT THIS BOOK PROVIDES

What does this book provide to you? It introduces many of the key elements about self-driving cars and does so with an AI based perspective. I weave together technical and non-technical aspects, readily going from being concerned about the cognitive capabilities of the driving task and how the technology is embodying this into self-driving cars, and in the next breath I discuss the societal and economic aspects.

They are all intertwined because that's the way reality is. You cannot separate out the technology per se, and instead must consider it within the milieu of what is being invented and innovated, and do so with a mindset towards the contemporary mores and culture that shape what we are doing and what we hope to do.

WHY THIS BOOK

I wrote this book to try and bring to the public view many aspects about self-driving cars that nobody seems to be discussing.

For business leaders that are either involved in making self-driving cars or that are going to leverage self-driving cars, I hope that this book will enlighten you as to the risks involved and ways in which you should be strategizing about how to deal with those risks.

For entrepreneurs, startups and other businesses that want to enter into the self-driving car market that is emerging, I hope this book sparks your interest in doing so, and provides some sense of what might be prudent to pursue.

For researchers that study self-driving cars, I hope this book spurs your interest in the risks and safety issues of self-driving cars, and also nudges you toward conducting research on those aspects.

For students in computer science or related disciplines, I hope this book will provide you with interesting and new ideas and material, for which you might conduct research or provide some career direction insights for you.

For AI companies and high-tech companies pursuing self-driving cars, this book will hopefully broaden your view beyond just the mere coding and development needed to make self-driving cars.

For all readers, I hope that you will find the material in this book to be stimulating. Some of it will be repetitive of things you already know. But I am pretty sure that you'll also find various eureka moments whereby you'll discover a new technique or approach that you had not earlier thought of. I am also betting that there will be material that forces you to rethink some of your current practices.

I am not saying you will suddenly have an epiphany and change what you are doing. I do think though that you will reconsider or perhaps revisit what you are doing.

For anyone choosing to use this book for teaching purposes, please take a look at my suggestions for doing so, as described in the Appendix. I have found the material handy in courses that I have taught, and likewise other faculty have told me that they have found the material handy, in some cases as extended readings and in other instances as a core part of their course (depending on the nature of the class).

In my writing for this book, I have tried carefully to blend both the practitioner and the academic styles of writing. It is not as dense as is typical academic journal writing, but at the same time offers depth by going into the nuances and trade-offs of various practices.

The word "deep" is in vogue today, meaning getting deeply into a subject or topic, and so is the word "unpack" which means to tease out the underlying aspects of a subject or topic. I have sought to offer material that addresses an issue or topic by going relatively deeply into it and make sure that it is well unpacked.

Finally, in any book about AI, it is difficult to use our everyday words without having some of them be misinterpreted. Specifically, it is easy to anthropomorphize AI. When I say that an AI system "knows" something, I do not want you to construe that the AI system has sentience and "knows" in the same way that humans do. They aren't that way, as yet. I have tried to use quotes around such words from time-to-time to emphasize that the words I am using should not be misinterpreted to ascribe true human intelligence to the AI systems that we know of today. If I used quotes around all such words, the book would be very difficult to read, and so I am doing so judiciously. Please keep that in mind as you read the material, thanks.

COMPANION BOOKS

If you find this material of interest, you might enjoy these too:

1. **"Introduction to Driverless Self-Driving Cars"** by Dr. Lance Eliot

2. **"Innovation and Thought Leadership on Self-Driving Driverless Cars"** by Dr. Lance Eliot

3. **"Advances in AI and Autonomous Vehicles: Cybernetic Self-Driving Cars"** by Dr. Lance Eliot

4. *"Self-Driving Cars: The Mother of All AI Projects"* by Dr. Lance Eliot

5. **"New Advances in AI Autonomous Driverless Self-Driving Cars"** by Dr. Lance Eliot

6. **"Autonomous Vehicle Driverless Self-Driving Cars and Artificial Intelligence"** by Dr. Lance Eliot and Michael B. Eliot

7. **"Transformative Artificial Intelligence Driverless Self-Driving Cars"** by Dr. Lance Eliot

8. **"Disruptive Artificial Intelligence and Driverless Self-Driving Cars"** by Dr. Lance Eliot

9. "State-of-the-Art AI Driverless Self-Driving Cars" by Dr. Lance Eliot

10. "Top Trends in AI Self-Driving Cars" by Dr. Lance Eliot

11. **"AI Innovations and Self-Driving Cars"** by Dr. Lance Eliot

12. **"Crucial Advances for AI Driverless Cars"** by Dr. Lance Eliot

13. **"Sociotechnical Insights and AI Driverless Cars"** by Dr. Lance Eliot.

14. **"Pioneering Advances for AI Driverless Cars"** by Dr. Lance Eliot

15. **"Leading Edge Trends for AI Driverless Cars"** by Dr. Lance Eliot

16. **"The Cutting Edge of AI Autonomous Cars"** by Dr. Lance Eliot

17. **"The Next Wave of AI Self-Driving Cars"** by Dr. Lance Eliot

All of the above books are available on Amazon and at other major global booksellers.

CHAPTER 1

ELIOT FRAMEWORK FOR AI SELF-DRIVING CARS

CHAPTER 1

ELIOT FRAMEWORK FOR
AI SELF-DRIVING CARS

This chapter is a core foundational aspect for understanding AI self-driving cars and I have used this same chapter in several of my other books to introduce the reader to essential elements of this field. Once you've read this chapter, you'll be prepared to read the rest of the material since the foundational essence of the components of autonomous AI driverless self-driving cars will have been established for you.

―――――――――

When I give presentations about self-driving cars and teach classes on the topic, I have found it helpful to provide a framework around which the various key elements of self-driving cars can be understood and organized (see diagram at the end of this chapter). The framework needs to be simple enough to convey the overarching elements, but at the same time not so simple that it belies the true complexity of self-driving cars. As such, I am going to describe the framework here and try to offer in a thousand words (or more!) what the framework diagram itself intends to portray.

The core elements on the diagram are numbered for ease of reference. The numbering does not suggest any kind of prioritization of the elements. Each element is crucial. Each element has a purpose, and otherwise would not be included in the framework. For some self-driving cars, a particular element might be more important or somehow distinguished in comparison to other self-driving cars.

You could even use the framework to rate a particular self-driving car, doing so by gauging how well it performs in each of the elements of the framework. I will describe each of the elements, one at a time. After doing so, I'll discuss aspects that illustrate how the elements interact and perform during the overall effort of a self-driving car.

At the Cybernetic Self-Driving Car Institute, we use the framework to keep track of what we are working on, and how we are developing software that fills in what is needed to achieve Level 5 self-driving cars.

D-01: Sensor Capture

Let's start with the one element that often gets the most attention in the press about self-driving cars, namely, the sensory devices for a self-driving car.

On the framework, the box labeled as D-01 indicates "Sensor Capture" and refers to the processes of the self-driving car that involve collecting data from the myriad of sensors that are used for a self-driving car. The types of devices typically involved are listed, such as the use of mono cameras, stereo cameras, LIDAR devices, radar systems, ultrasonic devices, GPS, IMU, and so on.

These devices are tasked with obtaining data about the status of the self-driving car and the world around it. Some of the devices are continually providing updates, while others of the devices await an indication by the self-driving car that the device is supposed to collect data. The data might be first transformed in some fashion by the device itself, or it might instead be fed directly into the sensor capture as raw data. At that point, it might be up to the sensor capture processes to do transformations on the data. This all varies depending upon the nature of the devices being used and how the devices were designed and developed.

D-02: Sensor Fusion

Imagine that your eyeballs receive visual images, your nose receives odors, your ears receive sounds, and in essence each of your distinct sensory devices is getting some form of input. The input befits the nature of the device. Likewise, for a self-driving car, the cameras provide visual images, the radar returns radar reflections, and so on.

Each device provides the data as befits what the device does.

At some point, using the analogy to humans, you need to merge together what your eyes see, what your nose smells, what your ears hear, and piece it all together into a larger sense of what the world is all about and what is happening around you. Sensor fusion is the action of taking the singular aspects from each of the devices and putting them together into a larger puzzle.

Sensor fusion is a tough task. There are some devices that might not be working at the time of the sensor capture. Or, there might some devices that are unable to report well what they have detected. Again, using a human analogy, suppose you are in a dark room and so your eyes cannot see much. At that point, you might need to rely more so on your ears and what you hear. The same is true for a self-driving car. If the cameras are obscured due to snow and sleet, it might be that the radar can provide a greater indication of what the external conditions consist of.

In the case of a self-driving car, there can be a plethora of such sensory devices. Each is reporting what it can. Each might have its difficulties. Each might have its limitations, such as how far ahead it can detect an object. All of these limitations need to be considered during the sensor fusion task.

D-03: Virtual World Model

For humans, we presumably keep in our minds a model of the world around us when we are driving a car. In your mind, you know that the car is going at say 60 miles per hour and that you are on a freeway. You have a model in your mind that your car is surrounded by other cars, and that there are lanes to the freeway. Your model is not only based on what you can see, hear, etc., but also what you know about the nature of the world. You know that at any moment that car ahead of you can smash on its brakes, or the car behind you can ram into your car, or that the truck in the next lane might swerve into your lane.

The AI of the self-driving car needs to have a virtual world model, which it then keeps updated with whatever it is receiving from the sensor fusion, which received its input from the sensor capture and the sensory devices.

D-04: System Action Plan

By having a virtual world model, the AI of the self-driving car is able to keep track of where the car is and what is happening around the car. In addition, the AI needs to determine what to do next. Should the self-driving car hit its brakes? Should the self-driving car stay in its lane or swerve into the lane to the left? Should the self-driving car accelerate or slow down?

A system action plan needs to be prepared by the AI of the self-driving car. The action plan specifies what actions should be taken. The actions need to pertain to the status of the virtual world model. Plus, the actions need to be realizable.

This realizability means that the AI cannot just assert that the self-driving car should suddenly sprout wings and fly. Instead, the AI must be bound by whatever the self-driving car can actually do, such as coming to a halt in a distance of X feet at a speed of Y miles per hour, rather than perhaps asserting that the self-driving car come to a halt in 0 feet as though it could instantaneously come to a stop while it is in motion.

D-05: Controls Activation

The system action plan is implemented by activating the controls of the car to act according to what the plan stipulates. This might mean that the accelerator control is commanded to increase the speed of the car. Or, the steering control is commanded to turn the steering wheel 30 degrees to the left or right.

One question arises as to whether or not the controls respond as they are commanded to do. In other words, suppose the AI has commanded the accelerator to increase, but for some reason it does not do so. Or, maybe it tries to do so, but the speed of the car does not increase. The controls activation feeds back into the virtual world model, and simultaneously the virtual world model is getting updated from the sensors, the sensor capture, and the sensor fusion. This allows the AI to ascertain what has taken place as a result of the controls being commanded to take some kind of action.

By the way, please keep in mind that though the diagram seems to have a linear progression to it, the reality is that these are all aspects of the self-driving car that are happening in parallel and simultaneously. The sensors are capturing data, meanwhile the sensor fusion is taking place, meanwhile the virtual model is being updated, meanwhile the system action plan is being formulated and reformulated, meanwhile the controls are being activated.

This is the same as a human being that is driving a car. They are eyeballing the road, meanwhile they are fusing in their mind the sights, sounds, etc., meanwhile their mind is updating their model of the world around them, meanwhile they are formulating an action plan of what to do, and meanwhile they are pushing their foot onto the pedals and steering the car. In the normal course of driving a car, you are doing all of these at once. I mention this so that when you look at the diagram, you will think of the boxes as processes that are all happening at the same time, and not as though only one happens and then the next.

They are shown diagrammatically in a simplistic manner to help comprehend what is taking place. You though should also realize that they are working in parallel and simultaneous with each other. This is a tough aspect in that the inter-element communications involve latency and other aspects that must be taken into account. There can be delays in one element updating and then sharing its latest status with other elements.

D-06: Automobile & CAN

Contemporary cars use various automotive electronics and a Controller Area Network (CAN) to serve as the components that underlie the driving aspects of a car. There are Electronic Control Units (ECU's) which control subsystems of the car, such as the engine, the brakes, the doors, the windows, and so on.

The elements D-01, D-02, D-03, D-04, D-05 are layered on top of the D-06, and must be aware of the nature of what the D-06 is able to do and not do.

D-07: In-Car Commands

Humans are going to be occupants in self-driving cars. In a Level 5 self-driving car, there must be some form of communication that takes place between the humans and the self-driving car. For example, I go into a self-driving car and tell it that I want to be driven over to Disneyland, and along the way I want to stop at In-and-Out Burger. The self-driving car now parses what I've said and tries to then establish a means to carry out my wishes.

In-car commands can happen at any time during a driving journey. Though my example was about an in-car command when I first got into my self-driving car, it could be that while the self-driving car is carrying out the journey that I change my mind. Perhaps after getting stuck in traffic, I tell the self-driving car to forget about getting the burgers and just head straight over to the theme park. The self-driving car needs to be alert to in-car commands throughout the journey.

D-08: VX2 Communications

We will ultimately have self-driving cars communicating with each other, doing so via V2V (Vehicle-to-Vehicle) communications. We will also have self-driving cars that communicate with the roadways and other aspects of the transportation infrastructure, doing so via V2I (Vehicle-to-Infrastructure).

The variety of ways in which a self-driving car will be communicating with other cars and infrastructure is being called V2X, whereby the letter X means whatever else we identify as something that a car should or would want to communicate with. The V2X communications will be taking place simultaneous with everything else on the diagram, and those other elements will need to incorporate whatever it gleans from those V2X communications.

D-09: Deep Learning

The use of Deep Learning permeates all other aspects of the self-driving car. The AI of the self-driving car will be using deep learning to do a better job at the systems action plan, and at the controls activation, and at the sensor fusion, and so on.

Currently, the use of artificial neural networks is the most prevalent form of deep learning. Based on large swaths of data, the neural networks attempt to "learn" from the data and therefore direct the efforts of the self-driving car accordingly.

D-10: Tactical AI

Tactical AI is the element of dealing with the moment-to-moment driving of the self-driving car. Is the self-driving car staying in its lane of the freeway? Is the car responding appropriately to the controls commands? Are the sensory devices working?

For human drivers, the tactical equivalent can be seen when you watch a novice driver such as a teenager that is first driving. They are focused on the mechanics of the driving task, keeping their eye on the road while also trying to properly control the car.

D-11: Strategic AI

The Strategic AI aspects of a self-driving car are dealing with the larger picture of what the self-driving car is trying to do. If I had asked that the self-driving car take me to Disneyland, there is an overall journey map that needs to be kept and maintained.

There is an interaction between the Strategic AI and the Tactical AI. The Strategic AI is wanting to keep on the mission of the driving, while the Tactical AI is focused on the particulars underway in the driving effort. If the Tactical AI seems to wander away from the overarching mission, the Strategic AI wants to see why and get things back on track. If the Tactical AI realizes that there is something amiss on the self-driving car, it needs to alert the Strategic AI accordingly and have an adjustment to the overarching mission that is underway.

D-12: Self-Aware AI

Very few of the self-driving cars being developed are including a Self-Aware AI element, which we at the Cybernetic Self-Driving Car Institute believe is crucial to Level 5 self-driving cars.

The Self-Aware AI element is intended to watch over itself, in the sense that the AI is making sure that the AI is working as intended. Suppose you had a human driving a car, and they were starting to drive erratically. Hopefully, their own self-awareness would make them realize they themselves are driving poorly, such as perhaps starting to fall asleep after having been driving for hours on end. If you had a passenger in the car, they might be able to alert the driver if the driver is starting to do something amiss. This is exactly what the Self-Aware

AI element tries to do, it becomes the overseer of the AI, and tries to detect when the AI has become faulty or confused, and then find ways to overcome the issue.

D-13: Economic

The economic aspects of a self-driving car are not per se a technology aspect of a self-driving car, but the economics do indeed impact the nature of a self-driving car. For example, the cost of outfitting a self-driving car with every kind of possible sensory device is prohibitive, and so choices need to be made about which devices are used. And, for those sensory devices chosen, whether they would have a full set of features or a more limited set of features.

We are going to have self-driving cars that are at the low-end of a consumer cost point, and others at the high-end of a consumer cost point. You cannot expect that the self-driving car at the low-end is going to be as robust as the one at the high-end. I realize that many of the self-driving car pundits are acting as though all self-driving cars will be the same, but they won't be. Just like anything else, we are going to have self-driving cars that have a range of capabilities. Some will be better than others. Some will be safer than others. This is the way of the real-world, and so we need to be thinking about the economics aspects when considering the nature of self-driving cars.

D-14: Societal

This component encompasses the societal aspects of AI which also impacts the technology of self-driving cars. For example, the famous Trolley Problem involves what choices should a self-driving car make when faced with life-and-death matters. If the self-driving car is about to either hit a child standing in the roadway, or instead ram into a tree at the side of the road and possibly kill the humans in the self-driving car, which choice should be made?

We need to keep in mind the societal aspects will underlie the AI of the self-driving car. Whether we are aware of it explicitly or not, the AI will have embedded into it various societal assumptions.

D-15: Innovation

I included the notion of innovation into the framework because we can anticipate that whatever a self-driving car consists of, it will continue to be innovated over time. The self-driving cars coming out in the next several years will undoubtedly be different and less innovative than the versions that come out in ten years hence, and so on.

Framework Overall

For those of you that want to learn about self-driving cars, you can potentially pick a particular element and become specialized in that aspect. Some engineers are focusing on the sensory devices. Some engineers focus on the controls activation. And so on. There are specialties in each of the elements.

Researchers are likewise specializing in various aspects. For example, there are researchers that are using Deep Learning to see how best it can be used for sensor fusion. There are other researchers that are using Deep Learning to derive good System Action Plans. Some are studying how to develop AI for the Strategic aspects of the driving task, while others are focused on the Tactical aspects.

A well-prepared all-around software developer that is involved in self-driving cars should be familiar with all of the elements, at least to the degree that they know what each element does. This is important since whatever piece of the pie that the software developer works on, they need to be knowledgeable about what the other elements are doing.

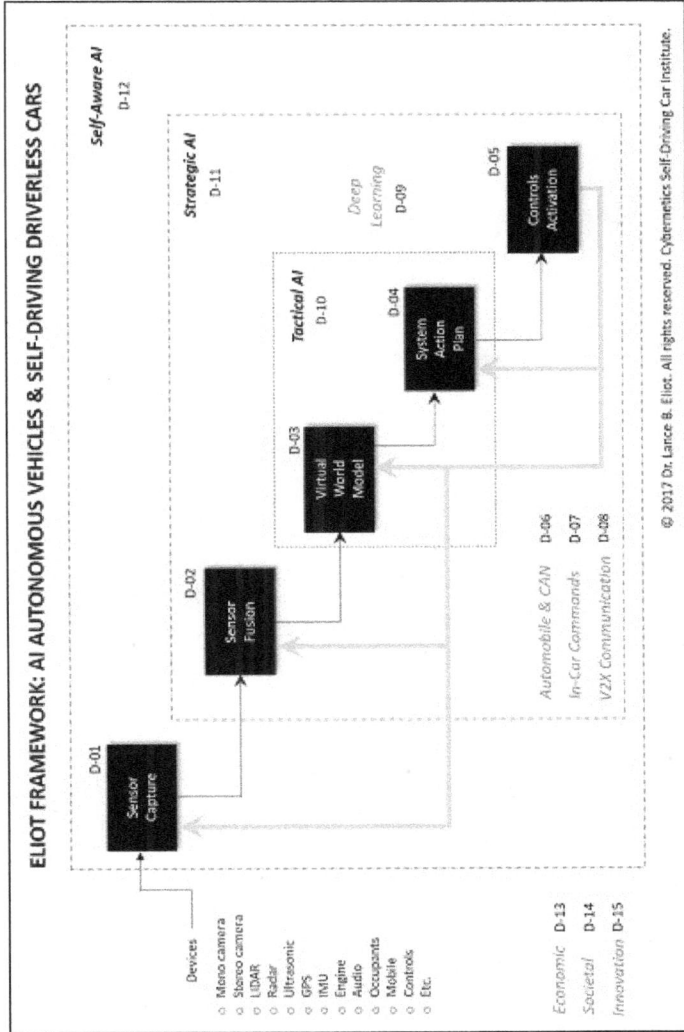

ELIOT FRAMEWORK: AI AUTONOMOUS VEHICLES & SELF-DRIVING DRIVERLESS CARS

CHAPTER 2

PRODUCTIVITY AND AI SELF-DRIVING CARS

CHAPTER 2

PRODUCTIVITY AND
AI SELF-DRIVING CARS

My daily commute to work takes about 1 ½ hours each way. Here in Southern California, my having a commute that is less than 2 hours each way is considered by some to be a blessing and I ought to relish the "lighter" commute than many of my colleagues. For those of you that live in other parts of the country, the idea of having a commute taking around 2 hours probably sounds crazy. Why would somebody drive that long each day, each way, just for a job?

Up in the Bay Area of Northern California, which I visit each month for work, my colleagues there have at times a similarly lengthy commute, but they tend to be on the BART (Bay Area Rapid Transit) trains during most of the commute time. I at first assumed that they would use that time for a boost in their productivity, being able to do so because they weren't having to drive a car as they made their commute. Turns out that being inside a train car does not necessarily lend itself to a much greater environment for trying to get things done. You are eyeing other people to make sure you stay safe, and you are contending with the noises in the train car of other people talking (and sometimes singing) and the movement and rattling of the train itself.

For my driving time in Southern California, I often schedule work related phone calls that I figure can be reasonably undertaken while in my car. This can be somewhat dicey though. I've found that callers will often realize that I am in my call while in my car, and at times be either insulted by the act or be concerned that they can be overheard. Admittedly, I was taken aback a few weeks ago when I was talking with a work colleague and he was in his car, and all of a sudden I heard his daughter speak-up and ask him a question, which, I was shocked since he and I were discussing rather sensitive matters at work and I assumed it was he alone in his car.

If I am taking a call during my drive time, I always let someone know if there's anyone else in the car with me. Turns out that my colleague felt no such need to do so, figuring that what the other person on the phone doesn't know won't hurt them. This seems a bit uncivil to me, but anyway, it happens.

Part of my rationale for taking calls while during my driving commute is that it seems to make me more productive. When I am solely focused on driving the car, and perhaps glancing at billboards off the freeway or admiring that sports car up ahead, I often feel like I am not being very productive. Sure, I try to entertain deep thoughts and solve tough problems, doing so in my mind. Many of my colleagues say that the commute driving time is the best time for them to try and figure out problems at work or at home, doing so with the silence in the car and the ability to put their minds to a gnawing issue.

I am sure that you might be concerned about people that take phone calls while driving their cars. We've all seen those drivers that seem quite distracted while apparently speaking to a non-visible person, presumably speaking to their phones and not just imagining that someone is in their car. Shouldn't that person be completely focused on driving instead? The effort to talk on the phone would certainly count as a distracting element and undermine the driving task.

Those that tend to use their phones while in the car, assuming they do so hands-free, would argue that there is no difference between talking with someone on the phone and talking with a passenger in the car. They would contend that if you are going to go berserk about a driver talking on the phone, you should likely go doubly berserk when a driver has a passenger in the car. The basis for being doubly berserk is that the driver might tend to turn their heads and look at the passenger, while with the phone approach the driver can continue to look straight ahead at the roadway. In that theory, the passenger is much more distracting to the driver than would be someone on the phone.

You could try to counter-argue that at least the passenger in the car might be able to aid in the driving task, perhaps serving as a second pair of eyes to be watching the road and maybe alert the driver to something amiss. The phone caller has no idea what's going on related to the driving of the car. Indeed, a passenger might realize when there are times to shut-up and let the driver concentrate, such as if there is an accident scene up ahead, while the caller on the phone might still be blabbering away and the driver is either too busy to let the caller know that things are hectic or doesn't want to tip their hand about the difficult driving situation.

I sometimes have several colleagues in my car at once, doing so during my commute or when we are going to a work meeting off-site. This then ups the ante, one might say, in terms of potential distraction for a driver. You have multiple passengers and they are likely to want to engage in dialogue about work or leisure or whatever. Should you also be engaged in those discussions? Well, you are certainly immersed there, sitting in the driver's seat, so how could you not participate? It would seem odd and perhaps rude to not participate in the discussions. Plus, if you seemed solely concentrating on driving, it makes some people nervous because they figure that if you can't handle some pleasantry conversations while driving, maybe you aren't suited to be doing the driving at all.

I've often wondered what would happen if there was a conflab of work people in a car and one of them suddenly and unexpectedly told the driver that he had just been fired from his job. Obviously, this might cause the driver to become erratic in their driving. Not a good situation. Hopefully this doesn't happen very often.

When my kids were young, I would drive them to school before I headed off to work. Sometimes they would try to catch a few extra winks, a bit of extra sleep, and take a kind of short nap on the way to school (it was only about a 15-minute drive). Or, they would try to finish up their homework. This was a questionable practice since they really didn't have much time to do so, figuring that the time needed to open their workbook, get focused on solving some homework assigned problem, figure out the answer, and right it down, well it was a lot to try and fit it all into a relatively short driving time.

The other problem with studying in the car involved the motion of the vehicle. At times, they would get motion sickness while looking down at their books and papers. This is a rough way to start a day and then end-up at school already feeling nauseous. Furthermore, the movement of the car would make it hard for them to write legibly on their homework papers. If you think that once they started doing their homework on tablet devices or their smartphones that this eliminated the issue of vehicle movement concerns, you'd be wrong. They had as much trouble trying to use the electronic keyboards as they did trying to write with a pen or pencil.

My overarching theme here is about how we use our time while in our cars.

What We Do in Our Cars

When you are driving a car, there is only so much else that you can accomplish other than actually driving the car. When you are a passenger, you can try to accomplish things and have a better chance since you presumably are not required to pay attention to the driving (assuming you trust the driver!). It can be hard though as a passenger to get a lot done since the rocking movement of the car tends to make it difficult to write or read, and you might also suffer from motion sickness while trying to do so. Taking a nap is maybe one handy way to use the time as a passenger, assuming that you are comfortable falling asleep in a moving car (some people aren't able to do so, either due to anxiety about wanting to be awake in case something goes awry, or due the movement of the car and potential nausea).

Some economists suggest that the time we spend in our cars today is relatively vacant of productivity. The normal passenger car driver is considered negligibly productive since they are focused on the driving task (a role that is perhaps only "productive" in terms of providing transit from point A to point B, but otherwise adds no further value, presumably). Passengers might have some amount of productivity, but it is considered rather low due to the nature of the "work" environment as available in a typical passenger car. All in all, the time we spend in our cars is often considered wasted or under-utilized with respect to being productive.

You can quibble with such a claim. If I am in the car with my work colleagues and we are discussing how to invent the next new mousetrap that will change the world, one could assert that the time in the car was very productive for all of us in the car. If I am in the car with my children while driving them to school, and we discuss the Constitution since they are studying it in their history class at school, I'd say this is productive time for both me and the kids. If I get a phone call from a colleague that is stuck trying to get his computer to function, and I am able to diagnose and offer a solution during the call, it seems to me that I was productive beyond just driving the car.

Overall, we need to try and grapple with the notion of productivity. What is considered as being productive and thus constituting productivity?

Trying to Figure Out Productivity

If you are measuring productivity of a person working on an assembly line in a manufacturing plant, you could in a straightforward manner count how many widgets they are making per hour. Then, if you can make changes to increase how many widgets that person is producing per hour, you've presumably increased the productivity of that person. It's the classic definition of productivity being the equation of output divided by input (for labor productivity, it is customary to use output volume produced as divided by labor input used).

Labor productivity in today's world is not so easily counted. We have been shifting to a knowledge-based form of workforce. You cannot quite so easily come up with simpleton measures for productivity.

I recall a Service Desk that I was asked to help boost the productivity of the workers or "agents" that were assigned to assist customers. The manager was under pressure to make those workers more productive. When I asked what the measure of productivity was, she indicated that it was the amount of time spent per call. Calls were usually about 10 minutes in length and the average agent was supposed to be doing about 5-6 per hour. The goal to become more productive was to drop the calls to 5 minutes in length and the hope was to double "productivity" to around 10-12 calls per hour.

I pointed out that one quick means to achieve such "productivity" would be to have the agents purposely short their calls. This consists of the agent opting to close-off a call in the shorter time period of 5 minutes, regardless of whether or not the agent had actually solved the customer issue. The customer would likely call again to try and finish the effort, which would be "beneficial" since it would further increase the call volume for the Service Desk. Problem solved!

But, of course, it doesn't solve the problem and instead creates new problems. The customer is bound to get irked by this trick. The customer will be steamed when they call back. The agent receiving the call will get the brunt of the anger. This will tend to undermine the performance of the agent. It will use up time for the agent to come up-to-speed about whatever was being discussed. And, there will be a number of customers that will opt to not call back at all and take their business elsewhere.

The gist of this example about the Service Desk is that we need to be very careful about how we define productivity.

A narrow definition such as the number of calls an agent takes per hour can be easy to come up with, but it then will lend itself to all sorts of abuses and undermine presumably what you are really wanting to achieve. I got the manager to also add other factors such as customer satisfaction and turn the productivity discussion into a more macroscopic look at the nature of the work being performed, the tasks being asked of the agents, whether tasks could be performed in other manners, whether the calls could be reduced or eliminated by taking other action that would prevent the customers from having to call anyway. And so on.

Are people today being productive in their cars? The answer to the question can differ dramatically depending on how you define productivity.

Applies to AI Self-Driving Cars

You might be wondering what does this have to do with AI self-driving cars?

At the Cybernetic AI Self-Driving Car Institute, we are developing AI software for self-driving cars. One aspect that many are hoping to have happen will be that people will become "more productive" in their cars due to the advent of AI self-driving cars. We think it is a topic worthwhile giving some pointed thought towards.

Allow me to elaborate.

I'd like to first clarify and introduce the notion that there are varying levels of AI self-driving cars. The topmost level is considered Level 5. A Level 5 self-driving car is one that is being driven by the AI and there is no human driver involved. For the design of Level 5 self-driving cars, the auto makers are even removing the gas pedal, brake pedal, and steering wheel, since those are contraptions used by human drivers. The Level 5 self-driving car is not being driven by a human and nor is there an expectation that a human driver will be present in the self-driving car. It's all on the shoulders of the AI to drive the car.

For self-driving cars less than a Level 5, there must be a human driver present in the car. The human driver is currently considered the responsible party for the acts of the car. The AI and the human driver are co-sharing the driving task. In spite of this co-sharing, the human is supposed to remain fully immersed into the driving task and be ready at all times to perform the driving task. I've repeatedly warned about the dangers of this co-sharing arrangement and predicted it will produce many untoward results.

Let's focus herein on the true Level 5 self-driving car. Much of the comments apply to the less than Level 5 self-driving cars too, but the fully autonomous AI self-driving car will receive the most attention in this discussion.

Here's the usual steps involved in the AI driving task:

- Sensor data collection and interpretation
- Sensor fusion
- Virtual world model updating
- AI action planning
- Car controls command issuance

Another key aspect of AI self-driving cars is that they will be driving on our roadways in the midst of human driven cars too. There are some pundits of AI self-driving cars that continually refer to a utopian world in which there are only AI self-driving cars on the public roads. Currently there are about 250+ million conventional cars in the United States alone, and those cars are not going to magically disappear or become true Level 5 AI self-driving cars overnight.

Indeed, the use of human driven cars will last for many years, likely many decades, and the advent of AI self-driving cars will occur while there are still human driven cars on the roads. This is a crucial point since this means that the AI of self-driving cars needs to be able to contend with not just other AI self-driving cars, but also contend with human driven cars. It is easy to envision a simplistic and rather unrealistic world in which all AI self-driving cars are politely interacting with each other and being civil about roadway interactions. That's not what is going to be happening for the foreseeable future. AI self-driving cars and human driven cars will need to be able to cope with each other.

Returning to the topic of productivity while inside a moving car, let's consider how things might change once we have truly autonomous cars at the Level 5. I do so to allow us to put aside the role of a human driver.

In the case of a true Level 5 self-driving car, the passengers should have no need whatsoever to be involved in or considering the driving of the car. As such, other than perhaps providing commands or conversing with the AI about various elements of the driving, the passengers aren't going to need to watch the road or look out for pedestrians or take on any other such driving chores.

When I say that the passengers might provide commands or converse with the AI about the driving, this would entail aspects such as telling the AI where you want to go. It might also include that during a driving journey you change your mind about your destination and thusly instruct the AI accordingly. Or, perhaps during the driving journey you get hungry and so ask the AI to take you to the nearest fast food restaurant. And so on. These though are aspects that have no need for the passenger to be engaged in the driving task itself.

You might wonder whether it is really possible that we will one day be in AI self-driving cars and not have to worry about the driving. In theory, indeed that is what is supposed to ultimately occur.

Right now, it is hard for most people to envision a future of that nature. As such, most of the surveys of people that ask whether they would be willing to be a passenger in a self-driving car are reporting a relatively high frequency of either no's or that the respondents would do so but then be warily eyeing everything about the driving task (therefore they would in a sense be "engaged" in the driving task, even if there aren't any driving controls, simply due to their concerns and lack of trust for the AI that's driving the car).

You might be surprised or shocked to know that I actually agree with the respondents that say they would be wary of being a passenger in a fully autonomous AI self-driving car, which makes sense to be wary because if one bases that view on what we have available today, I'd say they are right to be extremely cautious.

I realize you might ask people to think about the future, once we presumably will really have truly autonomous AI self-driving cars in abundance and routinely driving on our roadways. It is almost impossible to today put yourself into those future shoes.

Let's say some fifty years ago I took a survey and asked people whether they would be willing to carry around a device about the size of a pocketbook and they would be able to use it to call their friends, they would be able to electronically send their friends messages, they would be able to see all kinds of pictures and data from all-around the world on that device, etc.

I dare say that some would assume that such a device would be so expensive that it would be outside the affordability of any ordinary person. They would likely be hesitant to use such a device since they might fear that others could see or read what they are doing.

Imagine the kind of mental leap that would have been needed to envision a world that we have today, consisting of billions of smartphones all around the world and the Internet connections that we have.

I mention all of this because some have suggested that even if we have truly autonomous self-driving cars, some portion of people will be preoccupied with being worried about the driving and it will detract from any kind of "added" productivity that they might gain while inside such a self-driving car. I agree wholeheartedly that if the self-driving car at that juncture is not trustworthy to drive the car, it absolutely makes sense that people would be focused on the driving aspects.

People in The Future

Here's what I'd like to do herein.

Assume that there will indeed be a period of time for which people will rightfully be cautious and concerned when getting into a truly autonomous AI self-driving car. During that period of time, any productivity will be hampered by the aspect that the people in the car are having to double-check and watch over the AI. It would be like a passenger in a car being driven by a teenage novice driver, and you'd be nervous about every little thing the driver did or did not do.

Next, let's assume we get past that period of time and exit successfully from the jitters of trusting the AI to drive the car. Assume that we reach a point whereby you trust the AI as much as you would a seasoned human chauffeur that is expert at driving cars. In that case, by-and-large, for most people, you would be able to leave the driving to the AI, doing so without a sense of worry hanging over your head (similar to what most people do when in a limousine or even when in a bus).

Consider a perhaps handy analogy to airplanes. With airplanes, in the early days, most people were rightfully worried about getting onto an airplane. Over time, as airplanes improved and airplane travel became safer, most people have settled down and no longer have much of a concern when they fly. Sure, they still know that there's a chance of something going awry, and the flight attendants remind you of that potential during the pre-flight instructions, but I'd wager that most people are not especially thinking about the pilot and what the pilot is doing during their flights.

Does this imply that everyone is fine with flying? No, there are of course those that still have hesitations today about flying. The same would likely be true about AI self-driving cars. In that sense, no matter how safe and trustworthy the AI becomes at driving a car, there is still going to be a segment of society that has distrust to the degree that they will be unable to go into an AI self-driving car or will sit on pins during any such journey. I'm going to assume that this segment will be relatively small and it can be considered as a kind of rounding error with respect to people being willing to trust the AI and not be preoccupied with the driving task.

I want to then focus on the time period that involves no particular concern by passengers about the driving of the AI self-driving car. Meanwhile, if you want to argue with me about whether there will ever be such a day, and that maybe we won't ever get to truly autonomous AI self-driving cars, I'll offer the thought that yes, I agree it is possible we won't get there, but anyway for purposes of continuing this discussion, let's pretend or imagine that it will happen.

For this discussion, we'll henceforth assume that the passengers aren't concerned about the AI driving and those humans are able to devote the minds and attention to just about anything other than the driving.

Right now, any time a conventional car is being driven, it implies that there is a human driver in the car (they are doing the driving), and the human is generally occupied with driving the car. If we eliminate the need for a human driver, this means that anyone inside a car becomes a passenger and no longer a driver.

What might the former human driver that now is a passenger do while inside the AI self-driving car and while the AI self-driving car is on a driving journey?

Notice that I am purposely trying to make a distinction between those people that were already passengers in a conventional car and versus those people in an AI self-driving car that are now passengers and were once drivers. The former drivers are now able to do whatever a passenger can do, meaning no need to deal with the driving of the car. The passengers that were formerly passengers in a human driven car are now able to be passengers in the AI self-driving car, which means they are still passengers as they were before, but now doing so in a car being driven by the AI.

Don't get hung-up on this. You might say that someone that was formerly a driver might also at times have been a passenger. Yes, of course. I drive myself to work and often for lunch become a passenger in someone else's car when we drive to a nearby eatery. In that manner, in one day, I am both a driver at one point in time and a passenger in another point in time.

What I am trying to convey is that in the aggregate, we are going to be taking all of those former drivers that spent time driving, and they will in the truly autonomous self-driving car become passengers. This means that all of that time formerly used for driving is no longer being used to drive. This also means that those former drivers can do whatever a passenger can do.

Suppose then that all of those former drivers were to do almost anything "productive" while inside the AI self-driving car. Even if they did something productive for only say one minute, it would mean that if we said before they had zero productivity as a driver, we have now leaped tremendously into their being productive because they now have anything other than zero as their productivity.

Wow, we have mushroomed the productivity immensely, solely by eliminating the human driving task and now having the former human driver be able to provide attention to anything that might be considered productive.

Tricks of Productivity Counting

This is important to note because it shows the trickiness about wanting to predict whether we will be more productive while inside self-driving cars versus conventional cars. You can say it is a slam dunk that we will be, merely by shifting off the driving task and then putting any amount of time toward something considered productive. Note too that it could be that you are productive for just 10 seconds, rather than 1 minute, and you still have boosted productivity, especially if you add it up across the total number of drivers.

Per various stats by the Department of Transportation, there are an estimated 222+ million licensed drivers in the United States and it is claimed that we each spend about 17,600 minutes per year on-the-average driving (we'll say that's about 300 hours per year). This suggests that there are 222 million x 300 hours = 66,600 million hours per year in the United States for purposes of driving.

If you assume that driving is a zero-productivity task, this implies that in the United States alone we are "wasting" about 66,600 million hours per year of potential productivity.

If those same drivers opt to become passengers in the same manner as they are being drivers today (number of trips, length of trips, etc.), it suggests that we have a chance of turning the 66,600 million hours into some amount of productive time.

Suppose we are only able to turn 1% of that time into something productive, this means that we still have something on the order of 666 million hours of added productivity per year.

Outstanding! AI self-driving cars that are truly autonomous have a back-of-the-envelope calculation that shows we could boost American productivity by adding over 650 million hours of productive efforts at just a 1% use of their time toward something productive while inside the self-driving car.

If you take a leap of logic and say that those hours are worth at least the value of the national minimum wage (in terms of what people could earn per hour) and use a federal minimum wage of $7.25 per hour, you then have $4,712 millions worth of labor that could be added to the amount of national labor per year. I don't want to go off the deep-end on this and so let's just leave it there for the moment (I'm sure my economist colleagues are cringing!).

Of course, we're only so far considering the productivity of the former drivers. We should also consider the productivity of the former passengers (which, are still passengers, but albeit with a potential for added productivity time if you excise any time previously spent aiding the human driver of the car for undertaking the driving task).

Here's then what we seem to have:

- Former human driving and the hours of those human drivers, which as a driver was considered at a zero productivity, can become a huge boom to productivity by converting those drivers into passengers (or, the driving time into passenger time), and those now passengers can do nearly anything productive, even the littlest bit, and yet still cause leaps in aggregate productivity beyond the former zero.

- Former passengers in conventional cars, which already had some amount of productivity since they weren't doing the driving task, presumably can have at least that same productivity in the AI self-driving cars, and perhaps even more productivity due to no longer being a second pair of eyes for the driver.

Let's further pursue this matter of the amount of potential productivity involved.

Most pundits predict that AI self-driving cars will be running non-stop and be available whenever you might need a ride. This implies that the availability of using a car, in this case we are saying the truly autonomous Level 5 self-driving cars, allows people to potentially travel more often in a car than they did before. In that case, the amount of time that the former drivers are going to be in a car, and the amount of time that the former passengers were in a car, might all increase (this can be considered a form of "induced" demand).

Thus, whatever we might already consider the added productivity can be boosted even potentially higher because we might have a lot more "passenger" time in the future than we do today.

There is another angle on the productivity question. Will we be able to be more productive in our time outside of being inside a car, due to the advent of AI self-driving cars?

Notice that I've only focused on the productivity time while inside an AI self-driving car, but there are some that believe by having AI self-driving cars it will change other aspects of our lives and make us more productive beyond just the act of being inside a car. If you go along with that notion, please add more to the mounting amount of added productivity due to AI self-driving cars.

There are other sides to this coin, which I'll be addressing in a moment.

For example, you need to be considering that some say we might actually spend less time in the aggregate in our cars once we have AI self-driving cars. If right now you drive your son to baseball practice, you are using your driving time to do so. If you drive your daughter to her activities, once more you are consuming driving time. With the AI self-driving cars of Level 5, you won't need to go along at all. Thus, we would need to figure out what reduction this has in terms of the conversion of former human drivers that are not going to become passengers per se in circumstances whereby they before needed to drive for some other purpose that didn't actually need them to be present.

As with most things in life, there will be factors that will add-to and others that will subtract-from these endeavors.

Time is Not a Monolith

In terms of productivity, the discussion herein has suggested that you could have 300 hours per year of potentially productive time handed over to you as a former driver of conventional cars. That seems like a lot of time and could be presumably used for all sorts of nifty things, including perhaps learning a new skill to enhance your existing set of talents. Imagine taking an on-line course that streams into your AI self-driving car. This might not be simply a canned video course but instead a highly interactive class being taught on a MOOC (Massively Open Online Course) basis.

Unfortunately, trying to portray the time as one monolithic chunk of 300 hours is quite misleading. The reality is that it will be maybe 1 hour per day, roughly (I realize that would be 365 hours in a year but give me a break and let's just say 1 hour per day as an easy approximation).

Worse still, it really isn't fair to say it is 1 hour per day since the odds are that as a conventional driver you were making around 3 trips per day, and if that continues in the future, it means that each driving journey is only about 20 minutes in length.

The upshot is that the 300 hours is realistically a series of somewhat disconnected 20-minute segments.

I say disconnected because if you drive to work in the morning you've done one of the three 20 minute segments, then you work for say 8 hours, you then drive home and get another 20 minute segment, and maybe after getting home for a while then go out to do an errand and get the other 20 minute segment into your 1 hour per day of driving time. These 20-minute segments are not back-to-back. These are segments separated in time and likely your attention and awareness has changed between each segment.

We can even debate whether the full 20-minutes is realistic since there might be time to get settled into the self-driving car and time needed to engage whatever internal system you might use for taking an online class or doing any kind of Skype-like meetings. It could be that the usable time of the 20-minute segments is more akin to 10-15 minutes of actually dedicated and uninterrupted attention.

If you are taking a class while in your car, you would need to have the material divided into rather micro-chapters of 10 to 15 minutes. The YouTube kind of desirable lengths for videos. Even those micro-chapters might need to take a minute or two of the course time to bring you back up-to-speed as to wherever you last left-off.

Another perspective would be to suggest that instead of taking training, you would be performing work tasks of some kind. The work tasks would need to be framed into those 20-minute or less time segments that you are traveling in your car. I had mentioned earlier that I often make phone calls while driving. Those phone calls are usually relatively short and so I probably would get one or two calls during those 20-minutes. Plus, I could do the calls via a Skype-like visual camera since I would no longer need to be focused on the driving task.

I think we should ponder though what people could do for those short segments while in their self-driving cars, and hopefully be something that adds to their productivity. It might be a challenge to do so.

It could be some task that is repeated for each of those 20-minute segments in that they do the same kind of tasks for any of the times they are in the self-driving car, or it could be they do something differently depending on the nature of the timing of the 20-minute segments.

For example, in my case, I do work phone calls during my morning commute and afternoon commute. But if I drive to do an errand after work, I am a lot less likely to be doing a work-related phone call, partially because my fellow workers are also trying to enjoy their evenings too. What might I do with that evening errand's 20-minute driving segment?

You might also recall that I mentioned my belief that if you are having a call with someone they ought to know that it is private or "public" if someone else is in the car with you. Suppose that the advent of AI self-driving cars pushes us more so toward ridesharing and ride pooling, which means that you might be more likely to have someone else in your self-driving car during your commute than you do now. Will that perhaps inhibit or reduce the ability or desire to make calls while in the self-driving car?

Redesign of Car Interiors

One aspect that seems to be a relatively sure bet is that you might be more likely to do group meetings while inside the self-driving car. The design for future concept cars suggests that the seats in a true Level 5 self-driving car might be swiveled and allow for face-to-face conversations with the occupants. This redesign is aided by the fact that there is no longer a need to put a steering wheel and pedals at the front of the interior and nor dedicate a driver's seat for driving purposes.

We might also reasonably anticipate that network speeds will be much higher in the age of Level 5 self-driving cars, including the adoption of 5G and then later on 6G. This means that inside of the self-driving car you are likely to be able to carry on group meetings with others that might be scattered all across the globe. Some might be in their offices and others might be in their self-driving cars. I am not saying that the network will be glitch free and nor that it will be the Star Trek kind of reception, but at least it will be smoother and faster than it is today.

Will the potential redesign of future cars make it easier to do reading and writing while in a moving car? Recall that I had mentioned the difficulties my children had doing so in a conventional car, while trying to finish up their homework, sometimes at the last minute.

The jury is out still on this question. If you are trying to do conventional reading and writing while in a self-driving car, using pen and paper, I suppose the AI self-driving car is not going to make much of a difference in terms of allowing you to do a better job at reading and writing than a conventional car of today.

On the other hand, if the self-driving car has all sorts of touch-sensitive screens mounted throughout the interior, which is being predicted as part of the future redesign, and you are able to also use your voice for auto-transcription into writing, I suppose this new kind of interior arrangement might make it more palatable to do reading and writing (plus, we'll presumably be doing video watching more than we'll be doing "reading" of the kind we do today).

Another factor I mentioned earlier was that my children at times would get motion sickness while trying to study in the car. If people are going to try and use their time in Level 5 self-driving car to do studying or otherwise anything other than looking out the windows of the self-driving car, we need to consider whether the likelihood and frequency of motion sickness will rise.

Another consideration about trying to do something productive while inside a self-driving car brings up an issue that some of the auto makers are already anticipating, namely the dangers of loose objects that could fly throughout the self-driving car and strike someone, which could happen if the AI has to hit the brakes suddenly or make a quick maneuver.

Today's internal seating tends to prevent loose objects from flying around willy nilly. If we have swivel seats in Level 5 self-driving cars, it means that we are potentially facing each other. This suggests that any loose objects like your smartphone or your bottled water could become an airborne missile that hits another passenger in the face or torso.

The auto makers are struggling with how to protect passengers in general when inside these futuristic redesigned cars. Where will the air bags be? What kind of seat belts will be best? Should loose objects be hooked into some kind of bungee cords to prevent the objects from getting too far from you? I think we can all agree that people of the future will want to bring various loose items into their cars, and we cannot somehow say to them that they need to check their loose gadgets into the trunk.

I had mentioned earlier that my kids sometimes took a quick nap while in our car. For the redesign of future cars, it is believed that the swivel chairs might be shaped to allow you to lean back and sleep, or maybe even be able to remove the swivels and put in place sleeper "seats" when you know for sure that you want to catch some winks.

From a productivity viewpoint, I suppose you might argue that sleeping inside an AI self-driving car is putting us back toward a zero in terms of added productivity. But, look at that point in a different way. If I am able to get some sleep while in my AI self-driving car, maybe it makes me more productive when I get to work. Thus, if you are only counting productivity while inside the self-driving car, it might be unfair because I've done something seemingly unproductive inside the self-driving car that made me more productive outside of the self-driving car.

In spite of the seemingly apparent logic that we will gain productivity by the advent of AI self-driving cars, there are some that suggest we might actually have productivity loses due to AI self-driving cars.

Let's consider some of the points about the possibility of productivity leakage or loses.

If the interior of an AI self-driving car is ripe with touchscreens and other electronics, perhaps we will use any available time inside the self-driving car for purely entertainment purposes. Maybe we'll all be watching cat videos and make no effort to better ourselves with the added time that we will have to do something while being ferried by an AI self-driving car.

Perhaps the swivel seats will allow us to have greater comradery with our fellow persons but distract us from doing work.

When you were a driver of a car, you might have been in a more serious mood due to the somber nature of the driving task. This seriousness could have caused you to be work mindful and be thinking work related thoughts while commuting to the office. In contrast, while in a true AI self-driving car, you could have lost the edge to work and opt to just have fun or be a mental vegetable.

I'd guess that in spite of these potential productivity drains, and though it might lessen some of the productivity gains, it seems hard to imagine that a net effect would be an overall productivity loss. The argument that there will be productivity gains still seems relatively unfazed, though I suppose a bit dented.

Societal Work-Related Changes

When I've been mentioning productivity, it has been in the context of work-related productivity.

This will likely mean that if we all start doing work in our cars, and rather than it being something that is happenstance, suppose instead employers come to expect that you will do work while commuting in your car. If that's the case, it opens up other aspects such as whether you are officially on-the-clock during that time and whether you should be paid for it. In general, there are a slew of laws and regulations related to these kinds of working arrangements.

I'm not suggesting that working in your self-driving car will be a new idea and surprise anyone. Instead, I am merely pointing out that if today in conventional cars we have just a small portion of workers that get work done, imagine the volume and magnitude of work being done while in AI self-driving cars. This would raise tremendously the visibility of working in your cars. I think we could expect new regulations and other concerns that would surface once a large proportion of working society is doing this.

There is another aspect that I lightly touched upon earlier that I'd like to revisit, involving changes in our society due to the advent of AI self-driving cars. Some believe that we might end-up living much further from work as a result of the convenience of AI self-driving cars. Right now, you likely dread having to drive an hour and deal with the high pressures of snarled traffic. Imagine that you were in an AI self-driving car, oblivious to the traffic. And, you were able to work while in your AI self-driving car.

You might decide to live 2 hours or 3 hours away from work, knowing that you can sleep in your AI self-driving car on the way to work, and/or get your work started from your "mobile" office and thus not worry about getting to the office promptly. This again suggests that we might be spending more time in our cars, and maybe a lot more than we do today.

The statistic about drivers in the United States spending 300 hours per year driving is potentially misleading about what will happen in the future. Instead of begrudgingly making that 20-minute commute each-way today, you might welcome with open arms a 60-minute or more commute each-way, allowing you more options of where to live. The impact being that whereas before we were grappling with how to deal with productivity when cut into tiny 20-minute or so segments, it could be that the future will have much longer segments as people actively choose to go longer distances.

We also should consider other kinds of "productivity" besides work specific productivity. Maybe you will be a more productive citizen by using the time in your self-driving car to study up on issues of the day and be better prepared to vote in elections. Maybe you will be able to do more community work, volunteering to aid a non-profit while you are there in your self-driving car and have time to spare. These could have measurable impacts on society as a whole, regardless whether they relate to specifically performing a job that you might have.

Conclusion

With true autonomous AI self-driving cars, it could be the best of times or it could be the worst of times. Maybe we turn toward using our available time in self-driving cars to become a sweatshop and everyone must get onto the treadmill of work the moment they get into their self-driving car. That's a rather doomsday kind of view. Maybe we are able to use the time in the AI self-driving cars to make some additional money, maybe add more to society by volunteering, and possibly even get to know each other a bit better. I like that scenario a lot more.

When I was driving our car with the kids in it, I relished being with them to take them to school, but I also was trying to watch the road and make sure they got there safely. If I could have been with them and focused on just them, it would have been nice. I dreamed of the day when an AI self-driving car would allow me to do so.

CHAPTER 3

BLIND PEDESTRIANS AND AI SELF-DRIVING CARS

CHAPTER 3

BLIND PEDESTRIANS AND AI SELF-DRIVING CARS

I was driving along Ocean Boulevard and Pacific Coast Highway (PcH) in Malibu and Santa Monica, California when I came to a stop at a crosswalk that had no traffic signal. It was dicey that this major crosswalk has no traffic signal since it is commonly used by pedestrians that are trying to cross from the "inland" side of the street over to the ocean side of the street (putting them nearly onto the sand of the beach).

During the summer months it is a continual herd of pedestrians trying to make use of the crosswalk. Usually, it takes at least one or more daring souls to start the herd into crossing since the cars are zooming along and generally pretending to ignore the crosswalk and won't acknowledge the people that are wanting to make use of it (the pedestrians tend to bunch-up on the curb and when there is a sufficient number or when they are exasperated at waiting, they then in solidarity march into the crosswalk to get across the street).

I bring up this particular occasion because of something that caught my attention.

In the pack of pedestrians there was an elderly man with a white cane. He was moving the cane back-and-forth and tapping it on the ground. By glancing at his face, I'd guess that he was blind (though he hard dark and heavy sunglasses on, and I couldn't directly see his eyes), and he was opting to head over to the beach with the other pedestrians. Using this crosswalk would seem especially dangerous for him since it was not controlled by a traffic signal. He would pretty much need to rely on the judgment of the other pedestrians to know when to go ahead and enter into the crosswalk.

If he entered into the crosswalk on his own, I have my doubts whether the crazy drivers would honor the aspect that he was blind. It would seem self-evident by the use of his cane that he presumably was blind, but I am just saying that the SoCal drivers are so focused on themselves that they probably either would not notice that he was blind or (worse still) not care. I can imagine some drivers that would have kept going and possibly even cursed at him for boldly using the crosswalk on his own. His best bet would be to hope that the other bunched-up pedestrians made a wise choice to cross, and he would cross along with them.

By the way, I was tempted to call this the story of "The Old Blind Man and the Sea." Or, since I was watching him, maybe "The Old Blind Man and the See." My humblest apologies to Ernest Hemingway.

He was moving relatively slowly and generally was in the middle of the pact of pedestrians that were crossing the street. The flow of the pedestrians was faster than his walking speed and so many of the pedestrians were moving around him to proceed forward. They gave him some extra space inside the pack, especially since his cane was being aimed a foot or two beyond his width, alternating from left to right of his body.

It was probably handy that there were a lot of pedestrians crossing since otherwise I'd have gauged that he would have ended-up maybe half-way across and the rest of the pedestrians would have completed crossing by then. A large enough swell of pedestrians was sufficient to cover him throughout his crossing. If there were just a few pedestrians, they would have helped to likely curtail traffic for the first part of his crossing, but then he'd have been on his own. One can only hope or assume that if traffic had come to a halt to let the bulk of the pack across, they would have the patience to let him finish crossing (again, it's a toss-up here if that's what the drivers would actually do).

I'm sure you are wondering whether or not California has a driving regulation that says you are supposed to stop and let a blind pedestrian cross the street. Yes, we do. I'm sure you then are thinking, well, if it's the law, shouldn't this blind man have no qualms about crossing the street in the crosswalk, whenever he wishes, whether alone or with others? I certainly agree that the theory is that the drivers here would abide by the law, but I can tell you that the drivers here aren't necessarily that law abiding to begin with (they readily driver over speed limits, they drive excessively fast in school zones, etc.).

I'd earlier in my career had a chance to get a glimpse, as it were, about the world as perceived by a blind person. When I was a university professor, I taught a class on software engineering that involved a lot of programming. On the first day of one of my classes in the Fall semester, a student came up to me before class started and explained to me that he was legally blind. He said that he was telling me not because he expected any special treatment but due to giving me a heads-up about his status.

He indicated that I did not need to do anything other than what I normally do when teaching my class. I asked him if it would help that when I wrote on the whiteboard that I perhaps carefully say out loud what I am writing. He said this wasn't needed because he had a volunteer aid that went with him to his classes and sat next to him to let him know what was being written on the board. He also told me that for the exams, since I was likely to handout written exams, he would have the same aid there to tell him the questions. The aid would not interact with him about the answers and only tell him what the questions were. If I had any concerns about the exams and whether the aid might help him "cheat" in some manner, he said that there were official and trusted TA's (Teaching Assistants) that were ready to do the same as his "interpreter" for the written exams.

I was excited to have him take the class and it got me curious as to how he would do. Could he do the intense programming that was required for the course? I tried to imagine that my eyes were unavailable and could not fathom how I could do such programming. It seemed to me that he would have to somehow keep in his mind the various large passages of code that were involved and try to run through the lines of code purely mentally. I normally look at my computer screen and scroll back-and-forth at the code, using my eyes to refresh my mind and try to punch out the code.

Anyway, he did a wonderful job in the class. He spoke-up as other students did. He got the programming work done. He did quite well on the exams. I would not have known he was blind in terms of his classroom efforts, other than his having let me know before the class got underway. I did discover that his mental model of the programs was especially impressive, and I believe was more extensive than most of the other students in the class. I wasn't sure whether this was because he happened to have a greater mental capacity for it, or whether by necessity he had to keep it in his mind in this manner.

You might find of interest that there are an estimated 1-2 million legally blind people in the United States, and perhaps 7-8 million people all-told that have some kind of "visual disability" that renders them relatively blind (some prefer the phrase "visually impaired"). There is a somewhat strict definition for the phrase "legally blind" in that it means you need to have 20/200 or less vision. The states that have the most people with a visual disability or impairment include California, Texas, New York, and Florida.

There is also a large segment that is over the age of 65. This would seem logical in that as you get older the chances of having eye or vision issues is generally more likely than when you are younger. The elderly blind man that was using the crosswalk was somewhat "typical" in that part of Los Angeles as there are a significant number of retirees living in that area of town.

I had mentioned earlier that I was doubtful that many drivers here would have even noticed that he was blind. I'll call these inobservant drivers. The inobservant drivers aren't watching to see whether someone is blind. When I've asked such drivers why they don't pay attention to this, they often tell me that they don't want to be discriminatory towards someone that is blind, so why not just forget about looking for blind pedestrians and treat all pedestrians the same way.

I suppose you can somewhat understand their logic. But, at the same time, it seems to me that it belies the point that the driving code specially cautions that drivers should be extra careful when driving near to someone that is blind. It's the law. I don't have much belief that these inobservant drivers are really so concerned about being non-discriminatory and that instead it is a laziness and lack of care for others that really motivates their lack of attention.

According to the driving code, one aspect that inobservant drivers often do, and which can make life especially harder for a blind pedestrian, involves coming to a stop part-way into the crosswalk. You've likely done the same, whereby you meant to come to a stop before you reached the crosswalk line but didn't well-estimate the braking needed and so ended-up slightly protruding into the crosswalk. I'm not saying you've done this when a blind person was using the crosswalk per se, and only suggesting that it is an easy enough driving action to do.

People that are sighted can presumably readily see that your car is protruding into the crosswalk and walk around it. This can be troubling at times for even sighted pedestrians, particularly when there is a large number and it isn't easy for them to flow around the front of your car that is intruding into their crosswalk walking space. I've seen many drivers that realized they made a foe paw by entering into the crosswalk and so they try to back-up. Unfortunately, this then can create troubles for the car immediately behind them, and you suddenly have several cars all trying in a daisy chain way of backing up while having already come to a stop at a traffic signal or crosswalk (backing up because the driver at the front of the pack goofed-up).

A blind person might not so easily detect that a car is protruding into the crosswalk. They are right to assume that the crosswalk is supposed to be unencumbered. I'm sure that anyone that is blind and has used the crosswalks here would know over time that there is going to be a relatively high frequency of instances involving cars that have not stopped properly in-advance of the crosswalk. I was thinking that part of the reason the blind man was moving his cane back-and-forth somewhat widely was perhaps due to his wanting to detect that kind of circumstance.

Besides the inobservant drivers, there are also drivers that try maybe too hard to help a blind pedestrian. These eager beavers want to make sure that the blind person has ample room and also try to protect them from other drivers. It's a nice sentiment, but sometimes goes over-the-top.

For example, I've seen some drivers that opted to honk their horn to warn others that a blind person was crossing the street. The logic seems to be that by honking their horn it gets those inobservant drivers to become observant. I believe it is also assumed by the horn honking driver that it will let the blind person know that they are being acknowledged.

My understanding is that the horn honker is generally being misguided by their attempts to be thoughtful. The honking of the horn will likely startle the blind person and they might not have any clue as to why you are honking your horn. Furthermore, if the blind person is actively making use of their listening skills, doing so to try and make-up somewhat for the blindness, it is conceivable that the horn honking might be especially startling to their open ears. I also wonder whether other drivers will comprehend why the horn has been honked. Some might think that the horn honker is being rude and maybe it could spark some kind of road rage.

I've seen some drivers that get locked into a kind of stalemate chess move with a blind pedestrian. The driver decides to come to a stop, doing so several feet in-advance of the crosswalk. The blind person does not necessarily at first know that the car is even stopped, since it is so far from the crosswalk. The driver, though well intentioned, perhaps starts to get impatient that the blind person isn't readily taking the opportunity to cross, and so the driver starts to creep forward. But, the blind person that has by then detected that there might be a car there, now becomes bewildered by what the driver is trying to do. They then each wait for the other to make the next move.

Being a pedestrian already has a number of challenges. Try to imagine that you are blind and look around you the next time you go for a walk where there is traffic. It can become quite apparent as to how tricky and scary it could be to deal with the nutty drivers that don't care about blind pedestrians and likely too don't especially care about pedestrians overall. As mentioned, even drivers that do care about pedestrians can go over-board and create situations that though well-intended can end-up being confusing and potentially dangerous.

What does this have to do with AI self-driving cars?

At the Cybernetic AI Self-Driving Car Institute, we are developing AI software for self-driving cars. One aspect of robust AI software for driving is that it considers how to deal with pedestrians, including the case of pedestrians that are blind.

Allow me to elaborate.

I'd like to first clarify and introduce the notion that there are varying levels of AI self-driving cars. The topmost level is considered Level 5. A Level 5 self-driving car is one that is being driven by the AI and there is no human driver involved. For the design of Level 5 self-driving cars, the auto makers are even removing the gas pedal, brake pedal, and steering wheel, since those are contraptions used by human drivers. The Level 5 self-driving car is not being driven by a human and nor is there an expectation that a human driver will be present in the self-driving car. It's all on the shoulders of the AI to drive the car.

For self-driving cars less than a Level 5, there must be a human driver present in the car. The human driver is currently considered the responsible party for the acts of the car. The AI and the human driver are co-sharing the driving task. In spite of this co-sharing, the human is supposed to remain fully immersed into the driving task and be ready at all times to perform the driving task. I've repeatedly warned about the dangers of this co-sharing arrangement and predicted it will produce many untoward results.

Let's focus herein on the true Level 5 self-driving car. Much of the comments apply to the less than Level 5 self-driving cars too, but the fully autonomous AI self-driving car will receive the most attention in this discussion.

Here's the usual steps involved in the AI driving task:

- Sensor data collection and interpretation

- Sensor fusion

- Virtual world model updating

- AI action planning

- Car controls command issuance

Another key aspect of AI self-driving cars is that they will be driving on our roadways in the midst of human driven cars too. There are some pundits of AI self-driving cars that continually refer to a utopian world in which there are only AI self-driving cars on the public roads. Currently there are about 250+ million conventional cars in the United States alone, and those cars are not going to magically disappear or become true Level 5 AI self-driving cars overnight.

Indeed, the use of human driven cars will last for many years, likely many decades, and the advent of AI self-driving cars will occur while there are still human driven cars on the roads. This is a crucial point since this means that the AI of self-driving cars needs to be able to contend with not just other AI self-driving cars, but also contend with human driven cars. It is easy to envision a simplistic and rather unrealistic world in which all AI self-driving cars are politely interacting with each other and being civil about roadway interactions. That's not what is going to be happening for the foreseeable future. AI self-driving cars and human driven cars will need to be able to cope with each other.

Returning to the topic of blind pedestrians, most of the auto makers and tech firms would tend to say that this is an edge problem. An edge problem is one that is considered at the edge or corner of what you are otherwise trying to solve. You tend to delay dealing with edge problems. In the case of the auto makers and tech firms, they are primarily focused on getting the AI to drive a self-driving car such that it does the usual kind of things like drive properly and not hit people, but this often does not include the trickier situations that might be considered rare in their book.

Admittedly, the odds of an AI self-driving car coming upon a blind pedestrian is going to be overall somewhat rare, since it all depends on where the AI self-driving car is actually driving. If an AI self-driving car is being used in a state that has almost no blind people, the odds are that the AI is not going to encounter a blind pedestrian. If the AI self-driving car is driving in a state that has an abundance of blind people, but the AI is not driving in areas for which blind people are more likely to be found, once again the AI might not have much chance of encountering a blind pedestrian.

In that sense, the auto maker or tech firm would argue that worrying about driving aspects for coping with blind pedestrians is low on the priority list.

They would also tend to argue that if their AI self-driving car is already programmed to deal with pedestrians in general, there is really not a need to do anything special regarding blind pedestrians. In the minds of those AI developers, they assume that if the AI driving legally, such as stopping properly in-advance of a crosswalk, it is sufficient as a means of handling any situations involving a blind pedestrian.

I consider such AI developers to be rather myopic in their views. I think that they perhaps cannot put themselves into the shoes of the blind pedestrian to understand how the aspects of the AI can potentially help or hinder a blind pedestrian (I refer to such AI developers as being egocentric in their designs).

As per the driving code, a car is supposed to give the right-of-way to a pedestrian that is blind. How can you or the AI figure out that someone might be blind?

As a human, you would look to see if the person had a white cane, and/or possibly a guide dog. You might also do a kind of facial recognition, trying to look at the face and eyes of the person, if possible. You might also observe how the person walks and moves. These are all tell-tale clues about whether the pedestrian might be blind.

For today's sensory systems on AI self-driving cars, by-and-large the depth of analysis of pedestrians is relatively shallow. Pretty much the goal of most AI systems of today is to just determine that an "object" is a pedestrian versus being say a fire hydrant or street post. The pedestrian is treated internally in the AI system as a kind of stick figure. The stick figure is at this position on the street, so many feet away from the self-driving car. The stick figure is moving and going toward the AI self-driving car or moving away. And so on.

Trying to figure out if the pedestrian has a cane is not included in many of today's AI sensory data interpretation routines. It's a hard thing to spot. It is thin and might be hidden by being next to a person's body, thus the body of the person and the cane seem to be one blob. Once the cane starts moving, it becomes perhaps even harder to detect due to the often-rapid motion of moving back-and-forth and while it is in close proximity to the person's body.

Many of the sensory systems don't even try to computationally pick apart a group of pedestrians, such as a pack that is standing at the curb or walking across the street in a crosswalk. It would take a lot of processing cycles to figure out each person in the pack, including what is a person versus not a person, which arms and legs go with which person, their direction and pace, and the rest of those separation aspects is all computationally hard to calculate in quick time.

A human driver can usually glance at a scene and be able to recognize that there are say five people in a pack of people, including there are two women, two men, and a small boy. You can likely gauge their ages and something about them by how they look, how they dress, how they carry themselves, and so on. You likely are able to notice readily where they are looking, and you can often do a reasonably good job of predicting what they might do next. A pedestrian that is hunched forward and appears eager to cross the street is more likely to get onto your watchlist, meaning that you are anticipating they might jump the gun on the crosswalk crossing and go sooner than the others.

Many AI developers of self-driving cars would argue that this kind of processing is just not in the cards right now. They would wave their arms and say that it is arduous to develop AI to be able to deal with this kind of sensory data. The sensory data itself is not pristine and so you need to contend with lots of noise and partial images or other kind of data issues. Many would say that just let the Machine Learning (ML) deal with this, but that's not yet a silver bullet on these matters either.

Based on the toughness of being able to discern a blind pedestrian, and along with the presumed rarity, there isn't much effort going toward trying to craft specific capabilities for this purpose. This means that blind pedestrians need to do their best to be no different than other pedestrians, which, places the burden somewhat more so on their shoulders, when in fact we should be hoping for AI self-driving cars that can do a better job than human drivers in coping with blind pedestrians.

That's why we are working on AI components for this purpose.

One might even predict that there could be more difficulties about AI self-driving cars and blind pedestrians since it is anticipated that most AI self-driving cars are more likely to be Electrical Vehicles (EV's) rather than gasoline powered cars. This makes a difference in that an EV is often much quieter than a gasoline powered engine. You've probably had an EV sneak-up on you and got surprised that the car got so close to you without you realizing it.

Imagine EV's that come up to a crosswalk and come to a halt. With a gasoline powered engine, you have a likely greater chance of detecting where it is, doing so by listening for the motor. If you are blind, and you've honed your listening skills, you can likely pluck out of the cacophony of street noises the sound of an engine that is a few feet away from you. If AI self-driving cars are mainly EV's, it will mean that for blind pedestrians they have one less lowered chance of detecting the car than if it were a nosier gasoline powered car.

Does this imply that an AI self-driving car should do things like honk its horn to alert that a blind pedestrian is nearby? I don't think we want AI self-driving cars to do what some humans do about blind pedestrians, even if those humans think they are doing the right thing. No, honking is not likely a good idea.

For some EV's, the auto makers are generally adding features to be able to purposely make a small noise such as hum or chime to let people know that the EV is nearby, which is being done generally to let pedestrians know of the presence of an EV. That feature would likely be helpful too for the case of blind pedestrians.

One aspect for AI self-driving cars that might increase the chances of possibly getting near to or somehow entangled with blind pedestrians involves the advent of ridesharing via AI self-driving cars. It is widely predicted that AI self-driving cars will be extensively used for ridesharing. Imagine that you can at any time of the day and any day of the week be able to readily get an AI self-driving car to come and pick you up. No need to deal with human drivers.

This also means that those AI self-driving cars are going to be pulling up to the curb in all kinds of varying situations. Today, we seem to have human drivers that are ridesharing drivers that will stop anyplace at all to pick-up a passenger. Will AI self-driving cars do the same? I ask partially because it can be startling as a pedestrian that all of a sudden, a car is hugging the curb just inches from you as you are walking along on the sidewalk.

This brings up too a different topic that relates to blind pedestrians, namely if a blind pedestrian wants to take a ride in an AI self-driving car, will the self-driving car and the AI be able to accommodate this kind of pedestrian? You would hope so.

The AI will need to ensure that it converses with the passenger in a sufficient manner to keep them informed about the driving journey, perhaps more so than a sighted person. This can get tricky too as to where to drop-off a passenger, since a sighted person might readily look out the window and tell the AI that the spot chosen is not a good one, but a blind person is unlikely to be able to make that same kind of judgement due to lack of sight.

I've not yet mentioned the role of guide dogs in this matter. First, similar to the limitations of today's sensory systems to identify pedestrians, the capability of AI for identifying dogs and particularly guide dogs is very primitive and not at all akin to what a human driver might be able to discern. You might be surprised to know that there are actually aren't that many official guide dogs in the United States, perhaps 10,000 or so, and thus the odds are that a blind pedestrian will probably not have a guide dog and would be using a cane alone instead.

In any case, we'll eventually want the AI to be able to discern dogs and other pets that are on the sidewalks and in our roadways, along with being able to figure out whether the animal has a particular purpose. In the case of the blind pedestrian, the presence of a guide dog could further aid the AI in ascertaining that the pedestrian might be blind (of course, it could be a sighted person training the guide dog, etc.).

Another factor of the AI involves it contending with other drivers that have detected a blind pedestrian. I mentioned before that some human drivers will go to extraordinary lengths to deal with a blind pedestrian.

For example, suppose a human driver suddenly decides to back-up, because they intruded into a crosswalk. If there is an AI self-driving car directly behind that car, the AI needs to also back-up, if feasible.

As also mentioned, there will still be lots of human drivers on our roadways during the time that AI self-driving cars are emerging, and so this is another aspect that the AI needs to deal with about human drivers.

Currently, many of the auto makers and tech firms are treating the aspects of what is behind the self-driving car as relatively unimportant. The number and types of sensors at the back of a self-driving car are rather limited and slim. The amount of processing devoted to analyzing what is behind the AI self-driving car is also relatively slim.

I mention this aspect because when an AI self-driving car has to from time-to-time backup, it could be that a blind pedestrian has made their way around the back of the AI self-driving car.

The other day, when a car that had intruded into a crosswalk opted to back-up, the pedestrian decided to go behind the car, which as you might guess was a bit of an issue since you now had a pedestrian that had decided to go around the car, but the human driver had decided to back-up so as to presumably not have the pedestrian need to go around the car.

It was nearly an accident that occurred in front of my eyes.

AI self-driving cars also need to be able to make right turns with great care, in case a pedestrian opts to step off a curb early or might be standing just off the curb.

Likewise, making left turns needs to be done with care. A pedestrian that is perhaps in a crosswalk that involves where the self-driving car is trying to make a left turn might be unaware that the car is making a left turn.

The AI needs to be looking for pedestrians that are in crosswalk and maybe taking extra time to get through it.

Blind pedestrians are already considered by the driving code and laws to be warranted for special treatment by drivers. It is incumbent upon the auto makers and tech firms to make sure that AI self-driving cars not just treat blind pedestrians as any kind of pedestrian but include extra capabilities and care for detecting and ensuring that blind pedestrians do not inadvertently get hit or run over by the AI self-driving car.

Let's not turn a blind eye to that important challenge.

CHAPTER 4
FAIL-SAFE AI AND
AI SELF-DRIVING CARS

CHAPTER 4

FAIL-SAFE AI AND
AI SELF-DRIVING CARS

Fail-safe. If you've ever done any kind of real-time systems development, and especially involving systems that upon experiencing a failure or fault could harm someone, you are likely aware of the importance of designing and building the system to be fail-safe.

For many AI developers that have cut their teeth developing AI systems in university research labs, the need to have fail-safe AI real-time systems has not been a particularly high priority. Often, the AI system being built is considered relatively experimental and intended to tryout new ways of advancing AI techniques and Machine Learning (ML) algorithms. There is not much need or concern in those systems about ensuring a fail-safe AI capability.

For AI self-driving cars, the auto makers and tech firms are at times behind the eight ball in terms of devising AI that is fail-safe. With the rush towards getting AI self-driving cars onto the roadways, it has been assumed that the AI is going to work properly, and if it doesn't work properly that a human back-up driver in the vehicle will take over for the AI.

I've repeatedly cautioned and forewarned that using a human back-up driver as a form of "fail-safe" operation for the AI is just not sufficient. Human back-up drivers are apt to lose attention toward the driving task and not be fully engaged when needed. Also, human back-up drivers might not be aware that the AI is failing or faltering and therefore not realize they should be taking over the driving controls. Even if the human back-up driver somehow miraculously realizes or is alerted to take over the driving task, the human reaction time can be ineffective and the time for evasive action might already be past.

There's another reason too to be concerned about the need for fail-safe AI in self-driving cars, namely that the hope and desire is that AI self-driving cars will ultimately need no human intervention whatsoever in the driving task. As such, if there is not going to be any expectation of a human jumping in to save the day, the AI has to be the one that can entirely on its own save the day. This means that even if the AI itself falters or fails, it somehow has to stand itself back up and continue driving the car. I'll say more about this in a moment.

At the Cybernetic AI Self-Driving Car Institute, we are developing AI software for self-driving cars and are doing so with a proverbial "from the ground-up" fail-safe AI core tenent. It's a key foundational aspect, in our view.

I'd like to clarify and introduce the notion that there are varying levels of AI self-driving cars. The topmost level is considered Level 5. A Level 5 self-driving car is one that is being driven by the AI and there is no human driver involved. For the design of Level 5 self-driving cars, the auto makers are even removing the gas pedal, brake pedal, and steering wheel, since those are contraptions used by human drivers. The Level 5 self-driving car is not being driven by a human and nor is there an expectation that a human driver will be present in the self-driving car. It's all on the shoulders of the AI to drive the car.

For self-driving cars less than a Level 5, there must be a human driver present in the car. The human driver is currently considered the responsible party for the acts of the car. The AI and the human driver are co-sharing the driving task. In spite of this co-sharing, the human is supposed to remain fully immersed into the driving task and be ready at all times to perform the driving task. As suggested herein earlier, I've repeatedly warned about the dangers of this co-sharing and/or back-up driver arrangement and predicted it will produce many untoward results.

Let's focus herein on the true Level 5 self-driving car. Much of the comments apply to the less than Level 5 self-driving cars too, but the fully autonomous AI self-driving car will receive the most attention in this discussion.

Here's the usual steps involved in the AI driving task:

- Sensor data collection and interpretation

- Sensor fusion

- Virtual world model updating

- AI action planning

- Car controls command issuance

Another key aspect of AI self-driving cars is that they will be driving on our roadways in the midst of human driven cars too. There are some pundits of AI self-driving cars that continually refer to a utopian world in which there are only AI self-driving cars on the public roads. Currently there are about 250+ million conventional cars in the United States alone, and those cars are not going to magically disappear or become true Level 5 AI self-driving cars overnight.

Indeed, the use of human driven cars will last for many years, likely many decades, and the advent of AI self-driving cars will occur while there are still human driven cars on the roads. This is a crucial point since this means that the AI of self-driving cars needs to be able to contend with not just other AI self-driving cars, but also contend with human driven cars. It is easy to envision a simplistic and rather unrealistic world in which all AI self-driving cars are politely interacting with each other and being civil about roadway interactions. That's not what is going to be happening for the foreseeable future. AI self-driving cars and human driven cars will need to be able to cope with each other. Period.

Returning to the topic of fail-safe AI, let's consider the nature of what it means to be considered "fail-safe" and then take a look at how it can be achieved for AI systems. There is often confusion amongst laypeople as to the meaning of fail-safe overall. Admittedly, it is a rather loaded term and one that carries various connotations and baggage associated with it.

Let's start with the major misnomer about being fail-safe. Some falsely assume that a fail-safe system is one that will never fail in any respects whatsoever. Though you can indeed try to use that as a definition, it is not quite so applicable in the real-world of real-time systems.

If we were to take the term "fail-safe" as meaning that something can never ever fail, no matter what, it would suggest that it must be impossible for failure to ever occur. I dare say that it is hard to imagine any kind of relatively complex and useful system that you might build that could reach to such a lofty goal. Never failing is a pretty high bar.

I realize it will be disappointing to some of you that harbor such a definition when I then say that the manner in which fail-safe is more commonly considered encompasses an allowance for failure in some respects, but that the system is able to handle the failure in a manner that will reduce, cover for, or avert the adverse consequences of the failure. For those of you that are perfectionists and insist on never allowing any kind of failure, it's a hard pill to swallow that I am allowing "fail-safe" to mean that failure can occur. Sorry about that.

It might now seem like we are opening the floodgates by allowing that a fail-safe system can experience failure.

If we are going to allow failure as an option, we next need to consider the range and depth of possible failures. Suppose that a teeny tiny failure occurs in some subcomponent and yet it is not particularly material to anything else that the system is doing. As such, yes, a failure occurred, but it is buried so deeply and has nothing to do with the final outcome of the system, thus, the failure in a sense never saw the light of day. On the other hand, we might have a massive single failure or a collection of a multitude of severe errors, all of which produce a final outcome of disastrous and human harming results.

The notion underlying a fail-safe system is that it takes into account the potential for failure, including failures big and small, and including failures having adverse outcomes and those not, and as a core element of the design and how the system is constructed, it anticipates those potential failures along with ways to mitigate those failures.

Sometimes at conferences where I talk about fail-safe AI, I get the question that if one is designing an AI system based on expected failures then aren't you really engendering a mindset of a willingness to allow failure to occur within the system?

My answer is not really. Sure, you could claim that by talking about potential ways of failure that it will creep into the minds of those designing and building the system, and perhaps they then carry that over into how they design and build it, committing into the AI system various failures that otherwise would have never occurred to them.

But, I don't think any well managed AI development would get itself into that bind. Instead, I would argue that the opposite is actually worse, namely that if you don't bring up potential failures beforehand then the AI developers will blindly and inadvertently possibly fall into the trap of silently and unknowingly designing failures and building failures into the system. I am a proponent that fore-knowledge is better than not having thought things through.

Another question that I sometimes get involves whether or not you can truly anticipate all possible failures, and even if so, the effort to design and build the AI fail-safe system to cope with all such failures would seem to cause the design and development to mushroom beyond any manageable size.

First, I'll absolutely concur that an AI system designed and built with a fail-safe mindset is likely to take longer to design and build, potentially costing more too, in comparison to a slap-it together mindset that does not include systems fail-safe and related safety considerations.

Right now, most of the AI self-driving cars at a more advanced level are being pushed into experimental trials on our roadways and the lack of fail-safe aspects can be somewhat masked or hidden from view. If you have a team of AI developers that each day are tending to an AI self-driving car, along with the human back-up drivers, it can in a sense keep at bay any apparent AI system failures. A day-to-day problem in the AI system is quickly detected and dealt with. This is not what's going to happen once such AI self-driving cars are truly allowed into the wild and become abundant on our roadways.

I'd suggest that you need to consider the overall total-cost over the lifetime of the AI self-driving car and weigh that versus the near-term get-it-out-the-door approach. An AI self-driving car that has weak or insufficient fail-safe AI will likely end-up either causing more so harm to humans, via say crashing into a wall or ramming into other cars, and up the ante on the overall cost of having deployed such a system. I've predicted over and over that there are going to be product liability lawsuits against AI self-driving car auto makers and tech firms. The question will certainly be posed as to what they did to incorporate fail-safe AI into their self-driving car AI.

Overall, I'm saying that it indeed is likely to take longer at the front-end to include fail-safe AI, but that over time this will be worth it. Not only due to the lives saved, and not only to avoid product liability losses, but also as a means to try and gain public trust that AI self-driving cars are essentially safe. Those auto makers that go for the first-move advantage will potentially put a bad apple into the barrel that will then have the chance of curtailing or disrupting all efforts towards achieving AI self-driving cars.

Now, in terms of the other part of the point about possibly elongating an AI development effort by wanting to incorporate fail-safe aspects, the second part of that question involves the implication that by trying to identify in-advance all possible points of failure you are taking on a nearly impossible task.

It would certainly seem daunting to have to consider for an AI system composed of thousands upon thousands of interacting parts all the places that a failure might occur. The number of combinations and permutations of potential failures would become astronomical and it would seem unlikely you could consider them all.

Yes, indeed, you need to be reasonable and thoughtful when trying to address the entire population of potential failures. There are ways to categorize failures and assign probabilities to them. You need to have some fail-safe features to tackle the failures of the highest chances and/or the highest consequences. You then would have other fail-safe features to serve as a catch-all, aiming to cover things for failures that you either cannot necessarily well-anticipate or that might fall into the bucket of less risky or less severe kinds of failures.

You might be thinking that perhaps no AI self-driving car should be allowed onto the public roadways until it is fully tested in a lab.

This is a nice thought, but one that seems less viable in the real-world. One argument is that until an AI self-driving car is on the public roadways, it cannot be fully tested. There is only so much testing you can do via simulation. There is only so much testing that you can do via closed tracks that are built specifically for AI self-driving car testing. It is argued that the complexity of driving on public roadways can only in-the-end be ascertained by actually having the AI drive a car on public roadways.

There's another underlying aspect that adds to this dilemma. For many of the AI self-driving cars, the auto makers and tech firms are establishing the AI to be able to learn over time. Using various Machine Learning aspects, the AI self-driving car is intended to learn as it goes. This is rather different than most non-AI non-ML systems that pretty much do the same thing once they are released into production and actual use. If you have a system that does the same things over and over, you can test beforehand quite extensively. If you have a system that is changing as it is in use, the ability to test the system begins more problematic.

Indeed, one aspect about a fail-safe AI system is that assuming that the AI is incorporating Machine Learning, you have a moving target of what the failures might be. You could have perhaps anticipated certain kinds of failures when the AI system was at state A, prior to having "learned" and progressed to state B. Now, at the state B, it could be that failures of new kinds might be able to emerge. It is as though you have a creature that can morph itself over time. It had two heads and a tail before being released, and now it has three heads, two tails, and five tentacles. Trying to anticipate failures for something beyond the initial states that you knew about can be quite challenging.

Let's next shift our attention toward the design and building of a fail-safe AI system.

Take a look at Figure 1 (see next page).

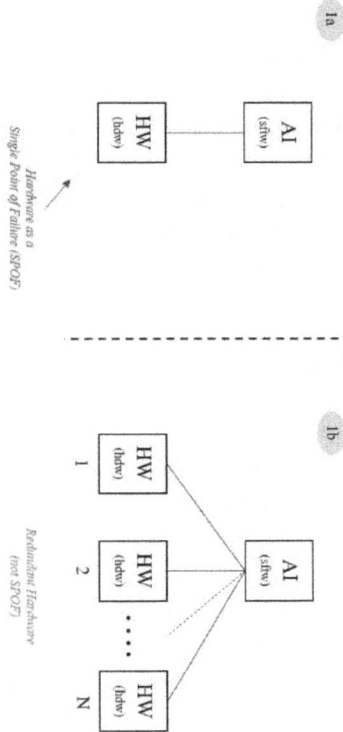

Eliot Framework: Fail-Safe AI via Redundant Hardware

Figure 1

Copyright © 2018, Dr. Lance B. Eliot, Cybernetic Self-Driving Car Institute.

In the diagram, I've divided the AI from the hardware and indicated that the AI is software and that the AI then runs on various hardware. This is an overall representation. I say that because I realize that some of you will start howling that the AI itself could be considered hardware and not just software. Yes, there are more and more AI systems that are incorporating specialized hardware that does chunks of the AI work, such as Artificial Neural Networks (ANN) that are embedded into GPU's (Graphical Processing Units).

For ease of simplification of this herein discussion, I'm going to go with the preponderance of things and say that the AI is software and it is running on hardware, and for which admittedly some of that hardware will be optimized for running the AI software. This point might not placate everyone, but I think it is a reasonable assumption for this discussion, thanks.

In Figure 1a, you can see that we have an AI system and it is running on some kind of hardware. The hardware is shown as one box on the diagram. By this depiction, I am saying that it is perhaps one or more processors and they are collectively bound toward running the AI software. Though it shows just one box on the diagram, it could be many processors bound together.

Whenever you are designing and building a real-time system, you ought to be looking around for any potential Single Point of Failure (SPOF) that your design or construction is engendering. For the moment, let's assume that the hardware that is running the AI system is absent of any redundancy. Thus, the AI software is dependent upon the hardware that is running the software, and if the hardware burps then it will cause the AI system to be disrupted.

As shown in Figure 1b, the easiest solution that most auto makers and tech firms are doing to deal with fail-safe AI is to add more hardware into the matter. We might have one or more hardware bundles, each of which is independently able to run the AI software. The AI could be running on all of them at once, and if any of the hardware bundles faults or falters, we would opt to make use of the other hardware.

In some less-than real-time systems, you might have hardware redundancy like this that consists of hot spares and cold spares. A cold spare implies that it is hardware that is not running the active software at the time and would need to spin-up to be engaged to run the AI software. A hot spare implies that the hardware is actively running the AI system and can be shifted to immediately.

Overall, this arrangement of the hardware, however you opt to do it, will certainly aid in eliminating a SPOF and will provide redundancy that should be able to keep the AI running in spite of more limited hardware failures.

This does not though eliminate any dangers or consequences that can arise from hardware failures. It could be that the switching time to have the AI shift into running on another available hardware bundle might cause a problem for the AI system during a real-time activity.

Imagine an AI self-driving car that is moving along on the freeway at 80 miles per hour and the AI is in the midst of trying to contend with difficult and complex traffic circumstances. Even a short delay in switching to another piece of hardware could be problematic.

Furthermore, suppose that the hardware that is on-board the self-driving car gets mushed all at once.

If the AI self-driving car gets hit from behind by a human drunken driver and imagine that we've secreted all the on-board processors including the redundant ones into the trunk of the self-driving car, it could wipe out the hardware entirely.

The AI software is not going to have much chance of doing anything once that hardware in-total goes out.

Generally, let's go with the idea that by adding the redundant hardware, we are aiding progress toward having fail-safe AI. We've got hardware that will generally be able to withstand more common kinds of physical failures such as a processor that goes bad or that its memory has issues, etc.

Take a look next at Figure 2 (see next page).

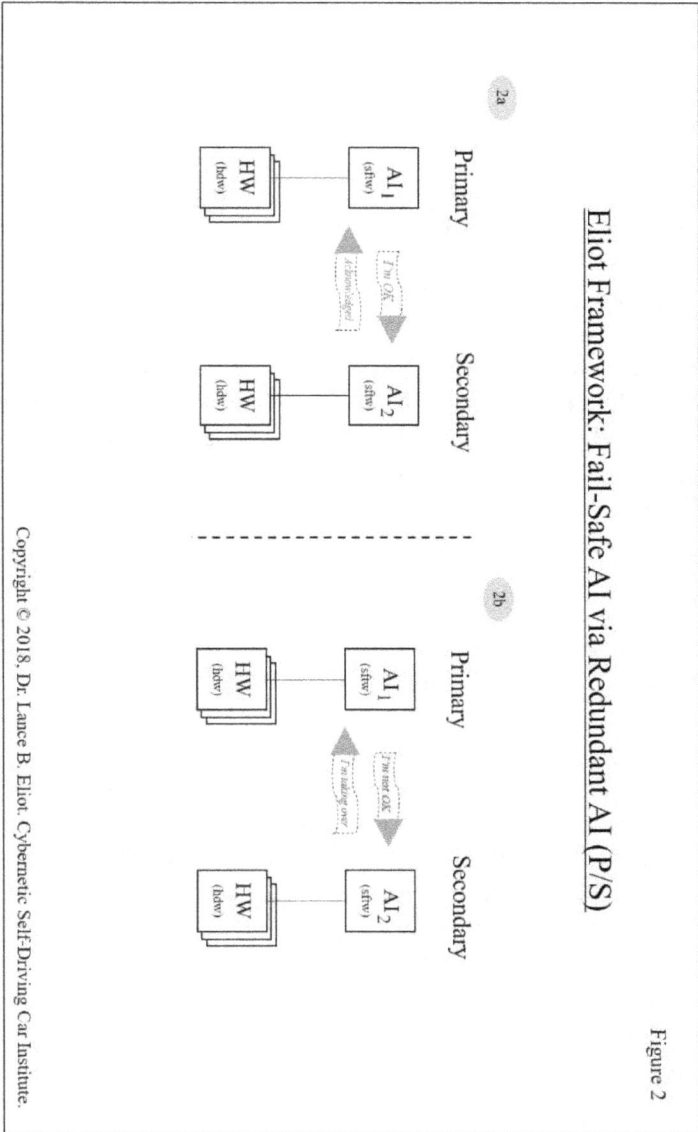

Eliot Framework: Fail-Safe AI via Redundant AI (P/S)

Figure 2

Though we are feeling good so far about the hardware redundancy, it still leaves us with the problem that the AI software itself is now the Single Point of Failure (SPOF). We need to do something about that hole.

As shown in Figure 2a, what we might do is setup another AI system and have it serving as a secondary for the primary AI system (I'll respectively refer to them as Primary AI and Secondary AI). We'll keep the redundant hardware underneath it all, and you should just assume that's the case for the rest of this discussion.

The Primary AI will periodically send a message to the Secondary AI and tell it that things are OK and there is not a need for the Secondary AI to step into the running of the AI self-driving car. The Secondary AI responds by acknowledging that it got the message from the Primary AI and that there is no need for the Secondary AI to engage actively into the driving of the self-driving car.

Per Figure 2b, at some point, the Primary AI might realize that things are amiss within itself and opt to instruct the Secondary AI to take over the driving of the self-driving car. I'll say more about this in a moment.

Generally, we would need to decide how often this messaging will take place between the Primary AI and the Secondary AI overall. It has to be frequent enough to ensure that the Secondary AI can step into the Primary AI role when needed and as soon as needed, but also not so frequent that it somehow becomes a nuisance or resource consumption beyond its worth.

Also, there would likely be an aspect that if the Primary AI fails to message to the Secondary AI that things are OK on the prescribed schedule, the Secondary AI might either try to make contact with the Primary AI or it might just opt to step into the driving task and take it over from the Primary AI (this would need to be carefully considered as to the downsides and upsides of such an "unrequested" step-in).

One key to this approach will be whether or not the Primary AI can realize that it is not OK. If the Primary AI assumes it is OK, but it really is not OK, it's not going to ask the Secondary to take over. The Primary AI should already have lots of designed and built-in capabilities for detecting its own status and be "self-aware" of what it is doing. Once it realizes that something is amiss, it would then presumably have the presence of "mind" that it should ask the Secondary to step-in.

Once the Secondary does step-in, whether by being asked or on its own (as mentioned earlier, if the Primary AI is non-communicative), we are now down to the bare bones. We now have a Single Point of Failure since we would assume that the Secondary AI is not going to ask the Primary AI to take over from it, and as such the whole ball-of-wax is now riding on the Secondary AI.

But, at least we do have the Secondary AI that is ready and willing and able to go. This does though bring up the aspect that the Secondary AI might have its own problems and not be able or willing to take over from the Primary AI. Let's consider this a bit further.

If the Secondary AI is say running on the hardware secreted in the trunk and the Primary AI is too, and if the rear-ender bashes up the hardware, perhaps the Primary AI would realize it is now degraded and so it asks the Secondary to take over. But, the Secondary might also now be degraded and thus refuse the request for it to take over. In that case, we might stick with the Primary AI and not resort to using the Secondary AI.

Take a look at Figure 3 (see next page).

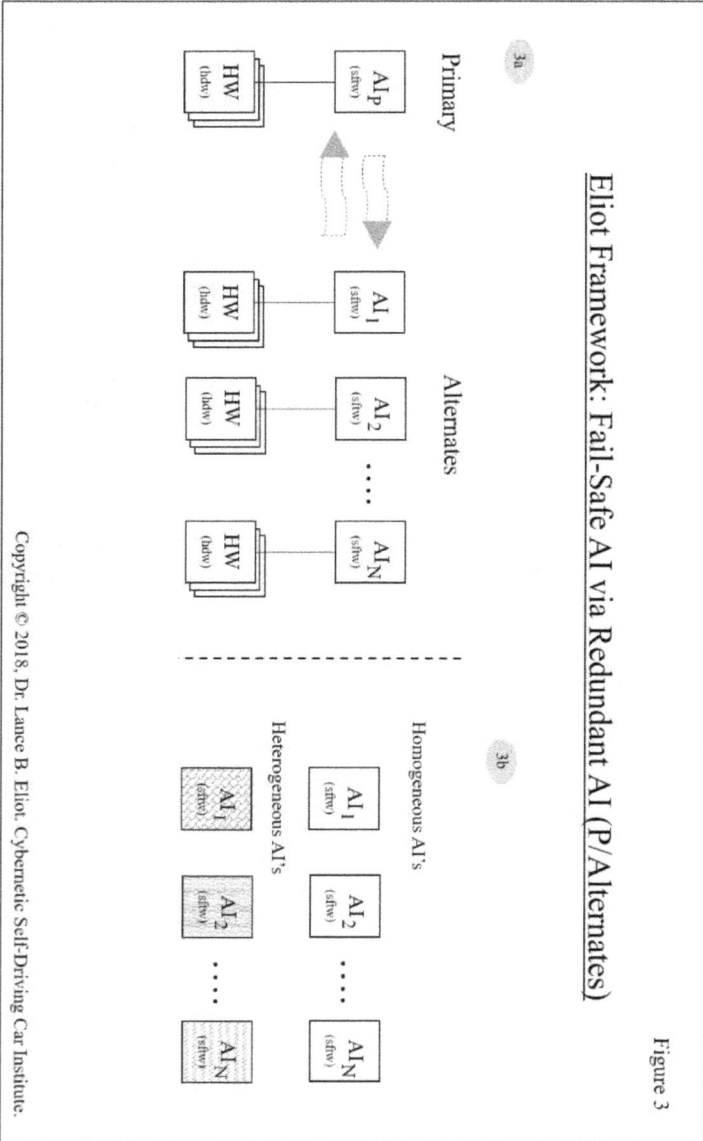

Eliot Framework: Fail-Safe AI via Redundant AI (P/Alternates)

Figure 3

There's another variation of this matter in entails how the Primary AI and the Secondary AI were designed and written.

Suppose that the Primary AI was designed and written such that we merely made a second copy and that became our Secondary AI. The question then arises that if the Primary AI has a fault that causes it to ask the Secondary AI to take over, will the Secondary AI potentially have the same exact fault since it is merely a carbon copy of the Primary AI? If that's the case, we aren't going to gain much in terms of having the Secondary AI (though, this can be somewhat ameliorated because it could be that the Secondary AI won't necessarily experience the same fault due to other exigencies such as the underlying hardware and other factors).

When we had just a Primary AI and a Secondary AI, it was earlier pointed out that once the Secondary AI takes over that it has no further recourse as to switching to anything else. It has become the sole cook in the kitchen.

As shown in Figure 3a, the Primary AI could potentially have a multitude of alternates, rather than being limited to just the singular Secondary AI. In that case of multiple Alternates, the Primary AI could potentially switch to one of them, which if it them faulted it could switch to the next, and so on. Or, we could have some other means by which the Alternates are chosen to be used.

Via Figure 3b, I show that it is also important to consider whether the AI alternates will be the same as the AI of the Primary, or whether you might have different AI systems that were designed and built separately from the Primary AI and from the other alternates. If you are using AI's that are the same, they are consider homogeneous alternates. If you are using AI's that differ from each other, they are considered heterogeneous alternates.

For the case of homogeneous AI's, you are really just copying the same AI and having multiple copies running or available. This is somewhat "easy" to do in that you aren't separately designing each of those AI systems and nor building them separately. This means that the odds of an in-common failure point is likely relatively high, since they were designed and built identically.

For the case of the heterogeneous AI's, you are separately designing and building the AI systems, doing so based on presumably the same set of requirements (otherwise, they are in a sense completely different AI systems). In the parlance of software engineering, this is often referred to as N-Version Programming (NVP).

N-Version Programming is a big step to take when designing and building a system, AI or otherwise.

NVP means that you are significantly increasing the effort and likely cost of designing and building the system overall. You are essentially doing two more complete development efforts. You only see this kind of approach in often very expensive and very complex systems such as for outer space rockets and such that you know will need to be self-sufficient on long journeys and often without any human control readily feasible due to the distances involved.

There are various thorny questions to be considered with N-Version Programming.

Should you use the same teams for each of the N versions? But, if so, you are likely to get the same outcomes. Should you have separate teams but allow them to share some of the code (such as if they are using open source that neither of the teams built themselves)? But, if so, you are likely carrying into each such "separate" system the same potential failure points. And so on.

This kind of approach is not for the faint of heart. You've got to be willing to go to bat for something that will be costlier and likely longer than otherwise to undertake, but at the same time hopefully provide a greater margin of fail-safe AI. You can argue about how much of a fail-safe you would get, and the trade-off between the added cost and effort versus the presumed added safety.

Take a look at Figure 4 (see next page).

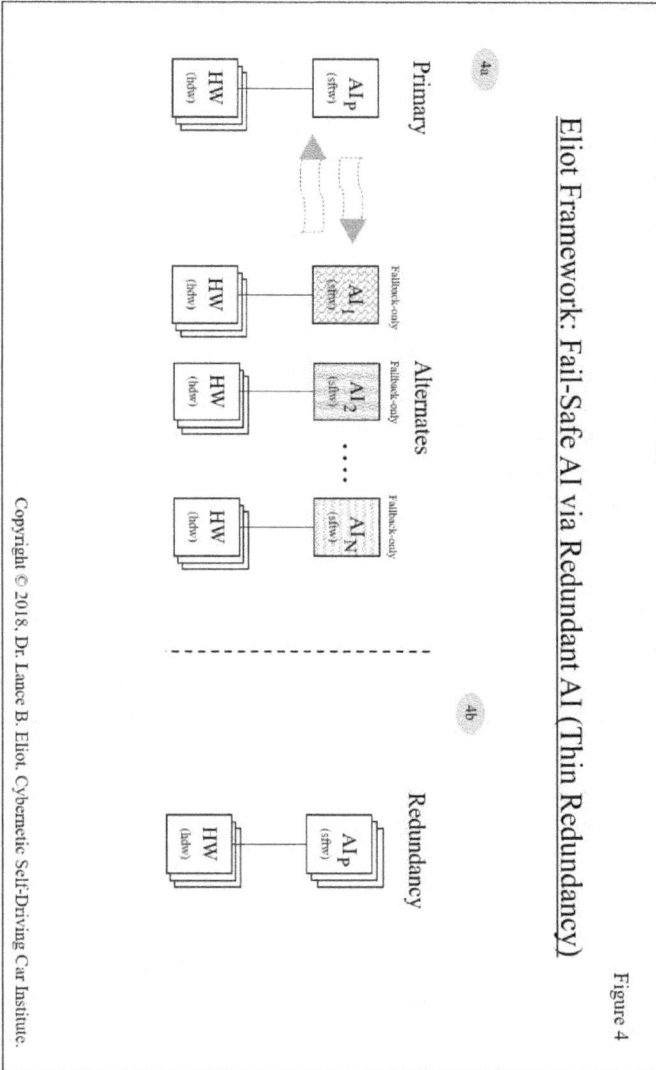

Eliot Framework: Fail-Safe AI via Redundant AI (Thin Redundancy)

Figure 4

As shown in 4a, the Primary AI has some N number of Alternates. So far, we've been assuming that the Secondary or the Alternates would consist of the same set of capabilities as the Primary AI, but that does not necessarily always need to be the case. Instead, it could be that the Secondary or Alternates are some subset of the Primary AI or some other aspect that pertains to the matter at hand.

In the case of an AI self-driving car, it could be that the Secondary or Alternates are only usable with respect to a fallback capability. If the Primary AI is having troubles, it would potentially be the case that the AI would want to perform a fallback operation and place the AI self-driving car into a minimum risk condition (such as parking safely on the side of the road). The Primary AI itself would likely have its own internal fallback capability, but if the internal fallback capability is faulty or the Primary AI otherwise becomes suspect, it might turn instead to one of the Alternates. Thus, it might be that the redundant AI is only with respect to the fallback operation.

The Primary AI might invoke an Alternate to undertake the fallback and if the chosen Alternate is unable to undertake the fallback it might then be handed to another Alternate. This would continue with each successive of the Alternates until hopefully one of them is able to actually perform a fallback. To try and avoid the chance that each of the fallback Alternates might have some in-common fault, they might be heterogeneous AI's that each were designed and developed independently.

When the redundant AI's are not equivalent in capability to the Primary AI, it is considered a thin form of redundancy since the redundancy only covers a portion of the full capabilities of the Primary AI. This also means that the Primary AI cannot hope to use the redundancy for the same range of capabilities as the Primary AI and instead would only switch over to a redundant AI when the circumstances were relevant to do so.

Figure 5

Eliot Framework: Fail-Safe AI via Redundant AI (Choice Redundancy)

5a

Redundancy

| HW |
| (hdw) |

| AI p |
| (sftw) |

Primary with Alternates

5b

Choice Redundancy

| AI a.l |
| (sftw) |

*All Alternates
(No Primary)*

| Chooser |
| (sftw/hdw) |

Potential SPOF

| HW |
| (hdw) |

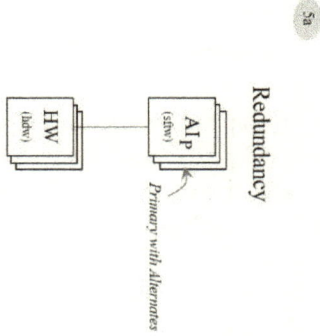

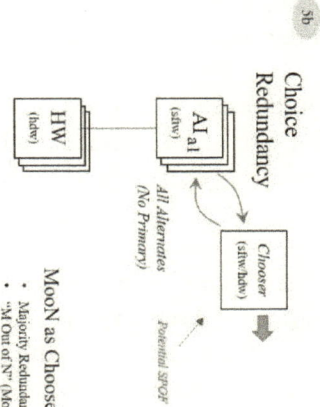

MooN as Chooser

- Majority Redundancy
- "M Out of N" (MooN)
- Choose via M agree out of N total
- Can be complicated, etc.

As was shown in Figure 4b, the format of suggesting that there are alternates to the Primary AI is indicated via the overlapping series of boxes behind the Primary AI, similar to the format being shown for the hardware redundancy.

In Figure 5, the 5a again indicates a Primary AI with its alternates.

A variant of the so far approach of having a Primary AI consists of not having any Primary AI per se and instead having a set of essentially all alternates. In that case, the question arises as to whom is calling the shots, so to speak, if there isn't a Primary AI doing so.

As shown in 5b, a system element labeled as the Chooser would be selecting among the alternates and determining which one is actively performing the driving task. This then would likely have all of the redundant AI's actively performing the driving task in a virtual sense, and the Chooser opting to select whichever one would seem best at the moment to be undertaking the actual driving task.

The Chooser would likely be relatively small and simple in its operation and mainly be trying to ascertain whether the Alternates are all doing OK or whether any are at a fault condition of one kind or another.

To help decide which if any of the alternates might be suffering a fault, the Chooser might not have any of its own logic per se to gauge the efficacy of what the alternates are producing and so instead might rely upon comparing the alternates.

For example, one sometimes usable approach is the M Out Of N (MooN) technique. If we have N number of alternates, we might beforehand determine that if M of them agree then those M are considered correct and the other N-M alternates are incorrect. This could be a simple majority and thus referred to as a Majority Redundancy.

Take a look at Figure 6 (see next page).

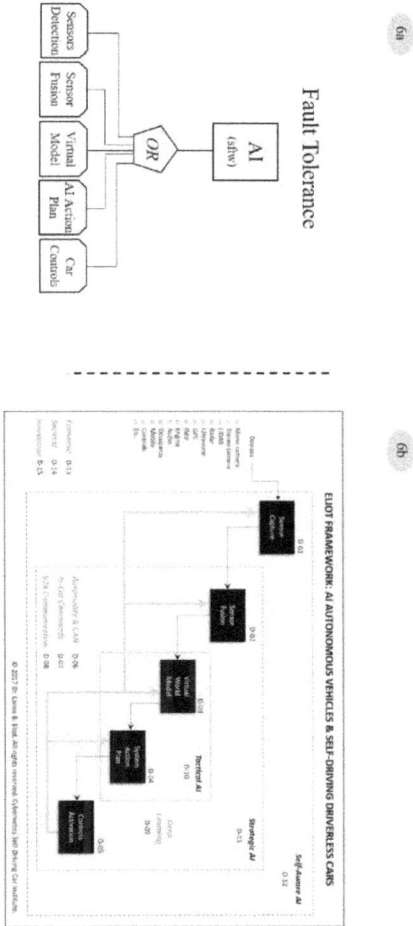

Figure 6

Eliot Framework: Fail-Safe AI via Redundant AI (Fault Tolerance)

In Figure 6, some of the subsystems associated with the Primary AI (or what might be a set of alternate AI's) is shown in 6a.

At the higher level of the AI system, the Primary AI is continually trying to gauge whether any of the subsystems might be encountering a fault or failure.

If so, this might either be handled internally within that AI instance or it might inspire the AI instance to then make use of an alternate.

Conclusion

Today's AI systems overall are generally quite primitive when it comes to being considered fail-safe or even at least fault tolerant.

This is partially due to the aspect that many of these emerging AI systems were developed in research environments that did not require the kind of resiliency that is needed for real-world real-time systems.

The danger for many of today's AI systems is that they are relatively fracture-critical, meaning that they are brittle and prone to falling apart, so to speak, upon the occurrence of even minor problems that might arise.

Unfortunately, even a modest fault or failure will tend to get the AI combobbled and have it produce an incorrect action or otherwise bring about an untoward result.

For AI self-driving cars, an incorrect action or untoward result can lead to a car that becomes a harm to humans in it and those near to it.

By considering the various means of using redundancy and other fail-safe AI techniques, an auto maker or tech firm that is designing and building the AI for a self-driving car has better odds at having robust and resilient AI. Robust and resilient AI can more readily deal with faults and either overcome those failures or at least be able to resolve a driving task with greater safety and aplomb.

CHAPTER 5
ANOMALY DETECTION
AND
AI SELF-DRIVING CARS

CHAPTER 5
ANOMALY DETECTION
AND AI SELF-DRIVING CARS

An anomaly is something considered out of the ordinary and often used to describe things or events that seem to be peculiar, rare, abnormal, or at times are otherwise difficult to even classify. Sometimes an anomaly is unwanted and can be bothersome when performing a task, while in other instances an anomaly might shed new insight that no one previously gave due attention to.

It can be hard to initially know whether an anomaly, once discovered, will be ultimately seen as desirable versus undesirable.

In the late 1800's, Wilhelm Roentgen was working in his lab on an experiment to make electrons zip through open air. After repeated trials with various cathode rays, he noticed that a screen at the edge of his table was lighting up (a barium platinocyanide covered screen became fluorescent). This was an oddity. It was peculiar. He could have shrugged it off as an anomaly that was not worthy of further attention.

Turns out that Wilhelm opted to study this aspect and it led him to the discovery of X-rays and X-ray beams. There were other researchers doing similar work at the time of his discovery, but it was his willingness to entertain the anomaly and give it its due that gets him credit for being the discoverer of X-rays or what some in-the-know refer to as Roentgan-rays. You've perhaps seen in the history books his first formal X-ray image, consisting of his wife's hand and depicted her finger bones and her wedding ring. One must say that she was quite brave to participate in the experiment, particularly since the nature of these electromagnetic waves and the hazards involved were not yet well understood.

Wilhelm's anomaly provides an example of a situation when detecting and acting upon an anomaly paid-off.

Sometimes an anomaly can be a fluke that provides no added value to what is being examined or studied. It could be random noise that happened to be encountered when you were doing something else and thus had no true bearing on the phenomena that you were studying. If you then pursue the anomaly so as to try and figure out whether it has merit or not, you might be wasting valuable attention and resources on something that has little or no benefit in the end. You are likely at first hopeful the anomaly will be a Eureka kind of moment, but often it turns out to be something mundane such as noise or a transient issue that was then self-corrected.

When I used to teach university classes on statistics and AI, I would cover the various "exclusion" techniques that could be used to deal with suspected anomalies. One obvious approach is to simply discard the anomaly. This though can create issues since it can leave a somewhat unexplained hole or gap in your research.

Another approach to deal with an anomaly in your data involves Winsorising it. Winsorising is a mathematical technique in which you substitute the anomaly with something from the nearest other data that is considered not an anomaly (referred to as non-suspected data). But this can be a questionable practice since it implies that you actually obtained "real data" that further supported your other data, when instead you essentially made-up or manufactured data to your liking. The same can be said for any other method used to substitute the actual data for concocted data.

One criticism of scientific studies and especially those in the medial domain are that at times the scientists performing a life-critical study will opt to toss out an anomaly that appears in their research. If you are trying to show that a new drug will save lives and prevent some dastardly malady from spreading, it can be tempting to disregard anomalies that might arise. By tossing out the anomaly or hiding it by a form of substitution, it could be that you are inadvertently hiding something that could be very telling. Perhaps the drug only works in certain situations and the anomaly could have revealed those crucial border-crossing aspects.

This is also why there is an ongoing clamor that researchers be willing to share their data when they publish the results of their work. By posting their data, it allows other researchers to examine the data and perhaps treat any anomalies differently than had the original researchers. This might reveal "censored" aspects of the study and thus open new questions about the conclusions of the research. In year's past, it was difficult to share data, but nowadays with the Internet there is not much of an excuse that it is arduous to do so.

You don't have to be a scientist to encounter anomalies. We experience anomalies in our daily lives. At times, you might not notice the anomaly, while in other cases you notice it but write it off as a fluke. In other cases, you might divert your attention to the anomaly, though this can be good or bad idea to shift your focus, depending upon the nature and value of the anomaly.

As an example, I was driving my car the other day on a lengthy driving journey and was using a major highway to do so. For hours on end, the traffic situation was relatively predictable and monotonous. It was a two-lane road in the northbound direction and passed through the California central region considered our state's agricultural belt. Regular cars would drive in the leftmost lane or "fast lane" and the lumbering trucks filled with various agricultural products such as oranges, onions, and so on, kept to the slower rightmost lane.

If a lumbering truck was going excessively slow, the other trucks behind it would try to go around the slowpoke truck and do so by briefly getting into the fast lane. Regular cars in the fast lane hated to have this happen. It meant that the fast lane, which was moving at 80+ miles per hour, would now need to slow down to allow a 55 miles per hour truck to proceed into the fast lane. Cars would either pretend to ignore the turn blinkers of the trucks that were trying to signal they wanted into the fast lane, or the drivers of some cars would blatantly prevent the trucks from getting into the fast lane by not allowing any gaps between the faster moving cars.

On the occasions that I opted to let trucks in front of me, I could see via my rearview mirror the pained expressions on the drivers that were in cars just behind me. They were exasperated that I was allowing the snail-paced trucks into the fast lane. How rude of me! Those other car drivers then rode on my bumper, trying to express their irksomeness and as though somehow it would pressure the trucks ahead of me to move along faster in the fast lane. Such is the civility of our roadways.

In any case, this was a routine matter and happened from time-to-time. Most of the time, nearly all of the time, the trucks were in the slow lane. I got used to passing truck after truck, all of them as though at a standstill in the slow lane, though it was just the perception based on the rapidly moving fast lane versus the much slower moving slow lane.

Towards the end of my drive, I saw up ahead that the trucks were in the fast lane. I figured that it was likely several trucks trying to pass a slower truck that must be turtle-like hampering the slow lane. I partially got into the slow lane to see what truck was causing the others to switch into the fast lane. To my surprise, there wasn't any truck at all up ahead in the slow lane.

This was curious. The trucks always loyally stay glued to being in the slow lane, unless there was a need to pass another truck, and in that case, they quickly got into the fast lane, went around the slower truck, and quickly got back into the slow lane. But, inexplicably, the trucks were all in the fast lane. They were still driving at the slower speed of 55 mph and yet all of them had decided to get into the fast lane.

An anomaly!

What should I have done? I could just stay in the fast lane, cruising at the slower 55 mph and followed the lead of the trucks. Or, I could switch entirely into the slow lane and zip ahead of the lengthy line of trucks in the fast lane. This is the reverse of what you might normally expect, in that usually you zip past via the fast lane, but if the trucks wanted to hog the fast lane, it seemed like they were nearly begging me to go ahead and use the slow lane (at least that's what I would have explained to a highway patrol officer that might have later stopped me for speeding past the trucks in the slow lane!).

I am sure that if there were other cars behind me at that point of the journey, they certainly would have been willing to use the slow lane for that purpose. Likely car after car would have come up to me, while I was lumbering in the fast lane behind the lumbering line of trucks, and then have gotten disturbed at the trucks being in the fast lane. Those car drivers would have probably gotten pissed off at the situation and then realized they could just skirt around the whole mess by using the slow lane. At that point, the slow lane would have become the fast lane, and the fast lane would have become the slow lane.

World turned upside down. I wondered whether those other car drivers would even take a moment to ponder why all of the trucks were in the fast lane. I'd bet that many of the drivers would not have given any thought to it. On these lengthy highway journeys, it seems like there are drivers that are just trying to maximize their speed and whatever seemingly legal or quasi-legal way to do so is fine with them. Not much thought involved. Almost a monkey-see and monkey-do kind of driving strategy. If there is a way to go faster, do so.

Well, I decided I would just stay in the "fast" lane and see how this matter evolves. After an agonizing five to ten minutes of being part of the lumbering herd (a distance of about 8 miles), it finally became apparent as to what was going on. Up ahead there was an accident in the slow lane and it was marked off with cones and flares. I am assuming that the truck drivers either spread the word amongst themselves or that somehow it had gotten marked onto a GPS mapping system, though mine did not seem to know about the accident. This was a rather remote location and so unlikely that anyone was helping to mark an accident that had just recently occurred.

The trucks had wisely gotten into the fast lane, in-advance of coming upon the accident scene. This was a good move since it avoided having to jockey and slow down at the accident scene itself. Instead, the trucks kept their regular 55 mph and were able to scoot past the accident. Once they had gotten reasonably past the accident scene, the trucks began to move back into the slow lane. World order was restored once again.

Let's revisit the story.

Why would I claim that this was an example of an anomaly? Due to the aspect that the trucks normally were in the slow lane and only briefly would get into the fast lane. If I had been collecting data during my driving journey, it would have plotted on a graph as showing that 99% of the time the trucks were in the slow lane and maybe 1% of the time used the fast lane (for passing purposes). This use of the fast lane in the case of the accident scene traversal was perhaps in that 1% of the time that the trucks were using the fast lane, but I logically could discern that the trucks weren't passing each other as was the custom for them to use the fast lane.

Suppose that I was an AI system that had been driving my car.

Would the AI have been able to discern that this seemed indeed to be an anomaly?

On the one hand, you might say that no, the AI would have not been able to do so. The trucks were legally in the fast lane and they had been using the fast lane from time-to-time, so nothing about this would at first glance seem odd or untoward. The AI would presumably not especially care that those were trucks ahead of it rather than regular cars. Sure, the traffic speed had slowed down but if the AI is doing a pied piper kind of approach of regulating its speed by the traffic ahead of it, the AI would just slow down the car and match the speeds of the trucks. No big deal.

Furthermore, the AI would want to drive "legally" (presumably) and so the idea of switching into the slow lane to pass the trucks would not likely have been something that the auto maker or tech firm had even included into the AI action plans for driving the car. Though using the slow lane for that purpose is not strictly illegal per se, it would be considered by many to be an improper or inappropriate driving tactic and so many AI systems would not consider it. I've debunked such ideas and have called for and predicted that AI for self-driving cars will need to be more flexible and not so narrow minded, as it were.

Overall, it would be likely that the AI of a self-driving car would probably not notice anything particularly unusual going on and would have simply stayed in the fast lane and followed along with the traffic. This is unfortunate in that it could be important for the AI to be watching for and possibly acting upon anomalies that it might encounter.

In my example of the trucks, you could argue that it makes no difference whether the AI was able to detect the anomaly of the traffic situation. Yes, luckily, in this particular circumstance, the AI-driven car being in the fast lane and staying in the fast lane was fine, and probably the more appropriate response to the anomaly. But, the aspect that the AI didn't even realize the occurrence of an anomaly and just blindly in a sense kept driving, that's the part that might not work out so well in other occasions of anomalies.

Here's another example that might better illustrate the matter.

I was on the freeway the other day and the traffic was light and moving along rather quickly (that's a rarity in of itself here in traffic snarled Southern California). I noticed up ahead that a man was walking along on the edge of the freeway.

Allow me to explain that most of our freeways here are relatively well blocked off from any pedestrians getting onto the freeway. There tend to be fences and brick walls that separate the freeways from any nearby homes, businesses, and so on. You would usually need to walk-up an on-ramp or exit-ramp of the freeway to physically be able to walk along the freeway. There are signs at the on-ramps and exit-ramps that clearly state to not walk onto the freeway.

The only time that you would normally see a person walking along the freeway would most likely be if their car broke down. They might then be walking to the nearest ramp so they could get off the freeway. But, this doesn't happen very often either since there are numerous specially dedicated phone boxes on the freeway that stranded people can use to call for assistance. Plus, our freeways are so frequently being cruised by police and tow trucks that the odds are you won't be stranded at your broken-down car for very long. And, it is generally publicized by the highway patrol that you should stay with your vehicle and not walk away from it (you can get a ticket for abandoning your car on the freeway).

Thus, the moment I saw a man walking on the freeway, I took a look at the side of the freeway to see if a car was broken down. I had not yet seen it and I looked up ahead and could not see one there either. This man, even if he was walking away from broken-down car, appeared to be quite a lengthy distance from his car. I right away doubted that this was a situation of a broken-down car.

When I first noticed the walking man, I was driving in the slow lane of traffic, which would have meant that I would end-up going past him shortly, doing so just a few feet next to this mysteriously curious and unusually encountered walking man. I would have gone past him at around 70 miles per hour, which is about 100 feet per second. I decided that the whole thing smelled and my innate Spiderman sense was tingling.

I moved over into the left most lane of the freeway, trying to create as much a separation of space between me and the walking man for when I would zip past him. I also was keeping my eye on him. I was on alert. My mind and my hands and feet were ready in case anything untoward might suddenly arise and I might need to maneuver my car rapidly.

I had in mind that the walking man might not be content with walking along the side of the freeway. Perhaps he might opt to suddenly dart into traffic. Or, maybe he might throw an object into traffic. Who knows? I realize you might be sympathetic to the walking man and think that maybe I was being a bit paranoid, but as I've tried to explain, it is highly unusual to see a person walking along the freeway and especially when there is no clear cut of indication of why he is doing so.

I would say his presence was an anomaly.

Should I have just ignored what I considered to be an anomaly? I opted to give the anomaly some credence. I took action by moving over to the fast lane and by keeping my eye on the matter. This action was not especially risky or odd, since I could have readily been in the fast lane anyway. It's not as though I suddenly slammed on my brakes or took any rash action. I took a subtle form of action that was intended to be a defensive form of driving and that would provide me with a lessened risk of exposure and a greater number of options if I needed to take other more pronounced actions.

What would AI do?

With today's AI, the odds are that the AI would likely have detected the walking man. The odds are that the detection would have led to the walking man being marked as such in the virtual world model that would be used by the AI to grasp the nature of the surroundings of the driving environment. The AI would certainly already be generally programmed for detecting and monitoring the movement of pedestrians.

Would the AI though have taken any action? Perhaps not. The pedestrian did not appear to be a threat to the AI self-driving car. He wasn't running into the lanes. He wasn't making wild motions. There was nothing obvious about any dangers associated with the pedestrian. If you didn't know any better, you would have classified the walking man as you would any person that might be walking on the sidewalk on any street that you might be driving on. In that sense, this seemed perfectly normal. At least it might seem so on the surface and without any deeper kind of assessment or analysis.

What made the hair stand on the back of my neck was the notion that a pedestrian was in a place and time that he should not have been, or at least that would rarely ever occur. And, I had already tried to determine whether this was a "normal" occurrence by looking for a disabled car (I say normal in the sense of from time-to-time, but rather rarely, there might be a walking person on the freeway due to the broken-down car aspects), but none seem to be anywhere nearby.

And so we now reach the crux of my theme, namely, as a human driver, I would classify this walking man as an anomaly. And, I would then consider whether to give merit to the anomaly or shrug it off.

Here was my thinking:

- If I shrugged it off, I would presumably continue unabated and pretty much ignore the anomaly.

- If I thought the anomaly had merit, I would investigate further, hoping to ascertain the validity of the anomaly. If the anomaly seemed to have sufficient validity, I would then decide upon whether my course of action should be altered knowing that I seem to have a genuine anomaly in-hand.

I assert that any well-qualified AI should be able to do the same, and especially for AI self-driving cars, which involve life-and-death kinds of matters and indeed that an anomaly can ascertain the fate of the humans in the self-driving car or nearby to the self-driving car.

At the Cybernetic AI Self-Driving Car Institute, we are developing AI software for self-driving cars. One important aspect of the AI is its capability to identify, detect, interpret, analyze, and determine a course of action related to anomalies.

Allow me to elaborate.

I'd like to first clarify and introduce the notion that there are varying levels of AI self-driving cars. The topmost level is considered Level 5. A Level 5 self-driving car is one that is being driven by the AI and there is no human driver involved. For the design of Level 5 self-driving cars, the auto makers are even removing the gas pedal, brake pedal, and steering wheel, since those are contraptions used by human drivers. The Level 5 self-driving car is not being driven by a human and nor is there an expectation that a human driver will be present in the self-driving car. It's all on the shoulders of the AI to drive the car.

For self-driving cars less than a Level 5, there must be a human driver present in the car. The human driver is currently considered the responsible party for the acts of the car. The AI and the human driver are co-sharing the driving task. In spite of this co-sharing, the human is supposed to remain fully immersed into the driving task and be ready at all times to perform the driving task. I've repeatedly warned about the dangers of this co-sharing arrangement and predicted it will produce many untoward results.

Let's focus herein on the true Level 5 self-driving car. Much of the comments apply to the less than Level 5 self-driving cars too, but the fully autonomous AI self-driving car will receive the most attention in this discussion.

Here's the usual steps involved in the AI driving task:
- Sensor data collection and interpretation
- Sensor fusion
- Virtual world model updating
- AI action planning
- Car controls command issuance

Another key aspect of AI self-driving cars is that they will be driving on our roadways in the midst of human driven cars too. There are some pundits of AI self-driving cars that continually refer to a utopian world in which there are only AI self-driving cars on the public roads. Currently there are about 250+ million conventional cars in the United States alone, and those cars are not going to magically disappear or become true Level 5 AI self-driving cars overnight.

Indeed, the use of human driven cars will last for many years, likely many decades, and the advent of AI self-driving cars will occur while there are still human driven cars on the roads. This is a crucial point since this means that the AI of self-driving cars needs to be able to contend with not just other AI self-driving cars, but also contend with human driven cars. It is easy to envision a simplistic and rather unrealistic world in which all AI self-driving cars are politely interacting with each other and being civil about roadway interactions. That's not what is going to be happening for the foreseeable future. AI self-driving cars and human driven cars will need to be able to cope with each other.

Returning to the topic of anomalies, the AI of a self-driving car has to be able to properly identify that an anomaly potentially exists, and so the first part of anomaly handling deals with detection.

The sensors of the self-driving car will likely already have various programs that examine the sensory collected data to try and find patterns. These include visual processing routines that handle the data collected via the cameras, encompassing both video and still images. There is software that does likewise for the radar, and for the ultrasonic sensors, and for the LIDAR (if so equipped), and so on.

Many of these pattern matching algorithms for examining the sensory data were likely trained via Machine Learning (ML). This gets us to the first area of concern about anomaly detection by a self-driving car. If the Machine Learning consisted of data that was scrubbed and had no anomalies, the appearance of anomaly out-of-the-blue during actual use of the system might go completely unnoticed. The sensory data interpretation programs might just shrug off the outlier data and consider it part of the noise and other transients that when is going to get when using sensors.

That's a tough aspect to overcome, namely, trying to figure out what is the usual kind of noise and transient data versus something that is a genuine anomaly worth considering. Suppose the AI was trained on all sorts of traffic signs, and then in the real-world a traffic sign that was not used in training is detected. The AI might opt to conclude that the traffic sign is not a traffic sign since it is outside of the pattern of what constitutes a traffic sign.

This happened to me in a human way when the other day there was a hand-written sign that a roadway crew had put up to forewarn about a hole or divot in the street up ahead. They tried to make it look like a regular traffic sign, but it was obvious to the human eye that it was a quickly crafted ad hoc sign.

What would the AI do about it? I would guess that the sensors would certainly have detected the presence of the sign. But, after trying to match it to the ones that it had learned from before, the odds are that it would be classified or categorized as just any kind of sign and not given its due related to the roadway and traffic situation (in contrast, for example, for the political elections, there are tons of signs put up all around town, none of which have anything to do with traffic, and thus it makes sense that a self-driving car would opt to ignore those signs).

The sensor data interpretation needs to be robust enough to give anomalies some attention, but at the same time if the anomaly is not relevant there is the issue of consuming the on-board processing cycles to try and ferret out the merits of the anomaly, which could perhaps starve some other crucial driving process. It is like a chess match that involves trying to determine how many levels deep, called ply, you want to do your analysis on. The deeper you consider the moves ahead in chess, the better the odds of making a good move now, but at the same time it chews up time and attention, which might be needed for other purposes (not so in a chess match, I realize, but this is so when driving a car).

Overall, there are some anomalies that genuinely do not exist but that the data or indication suggests it exists, and for which any pursuit is like going down a rabbit hole. There are other anomalies that have genuine origination and so need pursuit. One means to try and gain an indication of whether the anomaly has legs, so to speak, involves doing a kind of cross-triangulation on the anomaly.

In the case of sensor fusion, when the various sensory devices have provided their interpretations, it is up to the sensor fusion portion of the AI to aid in figuring out what might be a bona fide anomaly versus what might not be. By comparing the results of interpretations from each of the different sensors, the sensor fusion has the unenviable task of trying to figure out the real trust of what is surrounding the AI self-driving car.

Suppose the cameras have detected a shadowy image of something at the side of the road. The image is so hazy that it is not readily possible to classify the image as being a pedestrian versus being say a fire hydrant or a street post (or, maybe it is a false reading of some kind).

Meanwhile, suppose the radar has picked up a somewhat stronger set of signals and can present a more shaped outline of the object. And, let's suppose the LIDAR has done the same in terms of providing a clearer shape. By triangulating the multiple sensors, the sensor fusion might be able to discern that it is something that does exist and not just noise, and furthermore that it is pedestrian and not just an inanimate object.

The sensor fusion then passes this along to the virtual world model portion of the AI system. Within the virtual world model, there is now a numeric marker placed at the position of the suspected object in the overall model, and it is furthermore categorized with a probability that it is a pedestrian. The AI action planning program now examines the virtual world model to figure out what action, if any, the driving of the car should undertake given this news that there might be a pedestrian at the side of the road.

Here's the really tricky part that many AI systems are not yet considering. It is somewhat easy to consider the role of anomalies at the sensor data analysis aspects. The same can be said about detecting anomalies at the sensor fusion portion. It gets more complex once you are considering the virtual world model and the AI action planning portions.

Let's use my example about the walking man on the freeway.

I'm relatively confident that the AI self-driving car would be able to detect the walking man and determine that the object is a pedestrian. Sure, there could be issues trying to make this determination and it would depend on factors such as whether there is line-of-sight to the walking man for the sensors, and whether there is any weather that might be disrupting the sensor data such as rain or snow, etc.

Once the walking man gets placed into the virtual world model, would the AI realize that a walking pedestrian on the freeway is unusual? Would it be able to also extend that line of consideration and then look for other clues that might confirm the validity of the walking man being there, such as looking for a disabled car?

I'd dare say that most AI systems for self-driving cars would be unlikely at this time of being able to have that kind of anomaly seeking mindset.

In case you want to argue that the walking man was another example of no-harm no-foul in terms of if an AI system had not become concerned about the walking man (similar to my story about the trucks that got into the fast lane), I was waiting to tell you the end of the story about the walking man.

After I passed him, doing my 100 feet per second speed, and at a distance of about two lanes (let's say about 15-20 feet from him), he subsequently ran out into traffic. Many of the cars coming up were moving so fast that one of them ended-up striking him (I heard about it on the news, didn't see it happen directly). This happened in the slow lane.

I know that there are some AI pundits that will claim that had the AI self-driving car been in the slow lane it would not have hit the walking man because it would have miraculously made an evasive maneuver. I don't think it makes any sense to say that in this circumstance. The physics bely being able to avoid someone that suddenly darts in front of a car that is going 100 feet per second. You are just not going to be able to brake fast enough to avoid hitting that person. Where would you swerve to? Into other lanes of traffic? Or, maybe into the ditch next to the freeway, but perhaps kill the occupants of the car?

There are even some AI developers and AI pundits that would say that if a human was stupid enough to run into traffic, the person gets what they deserve. This is even a dumber thing to say. Suppose the car driver had swerved into the ditch and died, thus keeping the walking man alive. Is that a "deserved" death in the estimation of this idea that you get your just deserves? I think not.

Those same pundits might also argue that the walking man should not have gotten onto the freeway to begin with. As mentioned earlier, there are fences and brick walls that separate the freeway. Yes, it is possible to climb over those walls. Should we put up barbed wire and maybe gun posts, and make it seemingly impossible to get onto the freeway (a kind of modern-day Berlin Wall), doing so because the AI is insufficient to try and figure out when a pedestrian is there and should be avoided? I think not.

Those of us developing AI self-driving cars should be aiming to have the AI do the right kinds of actions, such as the action that I took, which I believe was a sound course of action. I had moved over into lanes away from the walking man and kept alert as to what the walking man was doing. There are other actions that could have possibly been done, such as maybe trying to block traffic and slow down traffic, or maybe call 911, but in any case, all of those actions rely on the realization that there was an anomaly afoot.

Robust AI for self-driving cars needs to give credence to anomalies. The AI needs to be overtly seeking out anomalies and giving them their due. This cannot though be done in a wanton fashion. There is only so much processing and bandwidth that the AI on-board system can undertake and do so on a timely basis. The AI needs to be watching out for false positives and not take action that is otherwise unwarranted and might carry its own risks. Nor should the AI be taken in by false negatives. Anomalies, love them or hate them, but either way you need to deal with them. That's the rub.

CHAPTER 6

RUNNING OUT OF GAS
AND
AI SELF-DRIVING CARS

CHAPTER 6

RUNNING OUT OF GAS

AND

AI SELF-DRIVING CARS

Running out of gas.

It's a pain in the neck.

During my college days, a friend of mine had an old jalopy of a car that was busted-up and yet it still managed to passably work, but the gas gauge seemed to have a mind of its own. I say this because sometimes the fuel gauge would be showing that the tank was half full, even though it was nearly empty, and at other times the gauge needle was on empty but there was actually a half of tank or more of gas in it. It was a wild guessing game as to how much fuel there really was in his car at any point in time.

Some affectionately referred to it as the Guess-o-Meter.

He carried in the trunk of his car a gas can filled with about one to two gallons of gasoline, doing so as a precaution in case he caught completely tricked by the gauge and ran out of gas while on-the-road. Everyone told him that carrying around spare gasoline in the trunk of his car was a nutty and highly dangerous practice. It wouldn't have taken much to have that gasoline get ignited, or perhaps if he got into a fender bender the gasoline might cause a small accident to turn into a torrent of fire.

You might assume that he should know how much gasoline he has in the tank of his car. All he would need to do is keep track of his fill-ups at the gas station and then track his mileage. By generally knowing how many miles per gallon his car was getting, he presumably could keep tabs on how much gas is likely left in the tank. Yes, this would have been a means to do a workaround regarding the broken gas gauge.

Unfortunately, he was the type of person that would not have been organized enough and systematic enough to actually do the kind of mathematical tracking that was suggested (I suppose the fact that he was allowing his car to continue to have a suspect gas gauge was a clear sign that he wasn't the kind of person that did things rigorously!).

We had a running gag that whenever you saw him, you would ask him how much gasoline he had left in his tank. I had thought he would tire of the joke and perhaps it would spur him to fix the gas gauge. Nope, it did not spur him to action. If anything, it seemed like he enjoyed the notoriety of being the guy that had a wild gas gauge. If the gauge didn't work at all, it would have been much less of a story. The aspect that it would willy nilly seem to decide how much gas was in the car made it much more entertaining as a story to be told.

I haven't yet mentioned the times that he did run out of gas. Most of the time that it happened, we never knew, since he would discretely make use of his spare gas stored in the trunk and then head to the nearest gas station to fill the tank. I have no idea how many times he did this. I'd bet that it was a regular occurrence and likely he was continually running out of gas and having to use his "fail-safe" operation to get back into business.

Many people don't realize that allowing your car to run out of gas can be bad in other ways, beyond just getting stranded someplace. The car doesn't especially like the notion of you letting the gas entirely be used up. The gas lines can get air in them, which can make it hard to restart the car once you've gotten more gas into the tank. Other things can go wrong with the engine by running out of gas. I know that my friend had to often tinker with doing various repairs under-the-hood and I suspect it was probably partially due to how he treated the car by letting it run out of gas.

He was so careless that he sometimes failed to re-fill his spare container of gas. In other words, he would use the gas can to provide gas once his car ran out of gas, he would then drive to the nearest gas station to fill-up the car, and he should have also then re-filled the gas can (I am not advocating the use of the gas can, I am merely saying that if that's what he was going to use as his back-up, he ought to make sure it was ready to be his back-up). Upon forgetting to re-fill his spare storage of gas, it then put him into the posture of having no back-up for when his car might run out of gas.

I was with him in his car on an occasion wherein this lack of being properly prepared arose.

He was driving along Pacific Coast Highway (PcH, as it's locally called), a scenic road in California that generally parallels the ocean. We ran out of gas. No problem, he insisted, and got out of the car to grab the spare cannister of gas in the trunk. I waited for him to quickly give us sufficient gas to then reach the next gas station. He got back into the car with a rather sad and sorrowful look on his face. Oops, the spare gas can was empty. One of us, he announced, would need to walk with the gas can to a gas station and bring back enough gas to get us underway again.

One of us? Wasn't it his responsibility to make sure that his car has sufficient gas? Believe it or not, he expressed that since we were both in his car, and both enjoying the use of his car, it really was the responsibility of both of us to tend to the car. Maybe that makes sense to you, but I assure you it made little sense to me at the time. I was perturbed.

In this case, we decided to push the car off the roadway and we luckily found a parking spot that we could have the car sit in without the threat of it being towed. We then both walked together to the nearest gas station. I suppose it is a memorable story now. At the time, I was irked the entire time of the adventure. From that point forward, whenever I got into his car, I would right away ask to check the spare gas can so that we would not get into a similar predicament again (I suppose that's my accepting the idea that we both had a joint duty or responsibility when both were in the car).

Even knowing that the spare gas can had enough gas for a quick trip to gas station was not always so comforting. We were driving through the Grape Vine, which is a somewhat mountainous pass that has a long and winding highway that gets quite steep at times, and I suddenly realized that there weren't any gas stations for miles upon miles. A road sign said that the next gas station was a great distance away. I pondered whether the spare gas would be sufficient to get us to the gas station, if we perchance ran out of gas.

When I mentioned to my friend that maybe we ought to reconsider our journey, he pointed out that we'd likely be Okay, especially if we reduced our gas consumption while in the Grape Vine. He proceeded to turn-off the Air Conditioning (AC). I'd like to mention herein that this was a hot summer day with outdoor temperatures ranging around 100 degrees or more. We began to swelter inside the car. He also turned-off the radio and anything else that he figured might use up gas.

We were nearly sitting on pins and needles as we traversed the Grape Vine. We kept the windows rolled-up because we had assumed that the aerodynamics of the car would be better if the windows were closed rather than opened. Given the heat outdoors and the mounting heat inside the car with its windows all closed, I wondered which would happen first, we would die of heat stroke or we would run out of gas.

He also offered a "clever" driving ploy (according to him). The Grape Vine has portions that go up, and portions that go down. You are gradually gaining altitude to make it through the mountains. You then lose the altitudes as you get past the crest and make your way back down towards the ground level. My friend pointed out that on the downslopes he would take his foot off the gas and we could coast. This would "for sure" reduce the amount of gas that we were using.

He also moved the car into the slow lane and tried to go around 35 to 45 miles per hour. He claimed that according to various gas mileage charts, the sweet spot for using the least amount of gas was around that speed. He said that when you drive at 55 miles per hour or faster, you are disproportionately using up gasoline. As you can guess, he had gradually learned all the ways to try and stretch out your gasoline use, though some of his ideas were borne of myth more than facts.

My children know that to this day I am a bit of a stickler about keeping gas in our cars. When the needle shows a quarter tank of gas left, I'm on the hunt to find a gas station. When they were learning to drive, I tried to instill this same approach into their driving style. They thought I was crazy. To them, let the needle get down to the level that the car tells you it is low. Most cars will provide a chime, or maybe a visual display or alert, and even let you know how many miles you can still go before running out of petrol.

I suppose it might have to do with the era of having a friend that had a Guess-o-Meter.

I grew-up becoming suspicious about the gas gauge. No sense in letting things get to close to the end, became my motto. Trust the car dashboard to take me up to the edge of the tank? No, thanks. I'll be attentive to my fuel needs and make sure to not reach the precipice, thanks.

According to statistics reported by the American Automotive Association (AAA), they respond to about 16 million or more calls per year for being out of fuel. Of course, that number is only accounting for those that have the AAA service. How many people per year really get stranded by running out of fuel? I'd bet it is an even bigger number.

When I refer to running out of gas, I am also alluding to running out of electrical charge too. I n other words, I suppose that I should be saying "running out of fuel" and thus you would know that I meant to include both gasoline powered cars and also electrically charged cars. I have friends that have Electrical Vehicles (EV) and even they sometimes say they are low on gas, even though they know and I know that it is not gasoline but instead electrical charge.

One of the drawbacks right now with EV's is the aspect of finding a place to charge your EV. Given the somewhat narrow range of miles that today's EV's can go on a single charge, you need to be mindful of where the charging stations are. Unlike gas stations that seem to sit on every corner, you aren't going to as likely be able find a place to charge your car. This is gradually changing as more charger locations are established, but for now, it can be a bit dicey to be using an EV for any kind of long-range driving.

Some of the tow services are now carrying with them a fast-charger to aid EV's that have run out of fuel. These nifty and life rescuing fast-chargers can potentially give your EV an added 10 miles of range by charging your car in about 10 minutes. I mention this aspect because another trade-off for some about EV's versus gasoline powered cars is that with gasoline you can in just a few moments fill your tank and be back on the road, while with most EV's you need to sit around for a while to let the charger get charged up.

Some EV's have a special slow-speed mode that allows you to stretch out the electrical power and thus make it to a charger someplace nearby. This feature is sometimes referred to as the turtle-mode or the crawl-to-a-charger mode. I suppose you can liken this to the tricks that my friend and I tried to play while on the Grape Vine, including turning off the AC, turning of the radio, keeping the windows rolled-up. Those same kinds of tricks can be used for an EV.

With some EV's, depending upon the model, you can potentially use the capability of regeneration to gain some added electrical charge. This generally takes the energy used by the brakes and returns it to some degree back into electrical storage bank. There are research efforts underway to have specialized tires that turn the roadway friction into electrical power for your EV. There are some that are also trying to put solar panels on the exterior of EV's, allowing for the catching of the sun's rays to charge the car. In one sense, you might say these are techniques akin to my friend's use of a spare gas can, though obviously much safer and sensible.

One of the extremely dangerous aspects about running out of fuel involves a car that is in-motion, which presumably is mainly when you would discover you are out of fuel. Without fuel to run the engine, your car now becomes a danger to those in it and those nearby to your car. You might be able to coast for a little while, but your overall ability to maneuver is now dramatically stinted.

The other day there was a driver on the freeway that appeared to run out of fuel. You could see his frantic look as he tried desperately to get his car over to the slow lane and then into the emergency lane. Freeway traffic had been moving along at a fast clip and so other cars were zooming past him, not wanting to be delayed or disrupted due to whatever crazy thing he was trying to do. These other drivers could care less about the driver that was frantically clawing his way to the side of the road. It was a quite dangerous situation.

One does have little sympathy for a driver that runs out of fuel. Did you not get a warning by your car that you were nearing the end of your fuel? Did you not take it seriously? Why didn't you sooner try to find a place to get fuel, rather than waiting until you actually completely were out of fuel? Don't you know how dangerous your coasting car can be? What kind of a person let's this happen?

Tow truck drivers often get the wildest stories from those that have gotten stranded and are out of fuel.

Some culprits will just admit they ignored the warnings. Some say they thought the miles left to go was enough to make it to a place to fill-up.

Some say their baby in the back-seat was crying and so they got distracted from the fuel gauge.

Some claim that the auto makers purposely have the gauge tell you one thing, such as that you have 10 miles left to go, but those auto makers actually know you have 20 miles left to go, and they do this to fool you into sooner getting re-fueled. I guess these people then figure that since they "know the trick" they can go ahead and ignore the 10 miles and assume it is more like 20 miles. It takes all kinds.

In any case, we have the grave danger of a car that has run out of fuel and that is in the midst of being on our roadways and has the potential for becoming an unguided missile. I say unguided and realize you might object and say that the coasting car should be able to be guided by the driver. Keep in mind that for some cars, trying to "guide" a moving car that has run out of fuel is not so easy. Also, you no longer can presumably use acceleration to get yourself out of a pickle.

From time-to-time, I have gotten into traffic snarls that involved a car that was stranded in the middle of the roadway.

Some of those instances have been cars that ran out of fuel. This brings up another aspect about the car that has run out of fuel, namely, whether the driver will be able to or even want to get out of the stream of traffic. I realize you probably assume that people would for sure try to get over to the side of the road. Apparently, some people are so terrified when they run out of gas that they just come to a stop in the middle of the roadway (or, it could also be that other cars would not let them get over to the side, so the driver figured they would just stay in the middle of the road).

These cars stranded in the middle of a roadway are absolutely a life-or-death danger. Other cars can ram directly into the stranded car. The occupants of the stranded car can get killed or injured, and the same can happen to the occupants of the car that hits the stranded car (plus, any residual crashes that occur as a result of the primary crash).

Being in the middle of the road, the occupants typically cannot get out of the car, for fear of getting hit by nearby moving cars. They instead sit there, a target, waiting for some inobservant driver to plow into them. I'll point out too that just because you can get your car to the side of the road, if feasible, it still does not mean you are safe per se. There are often cars sitting at the side of the road that get hit by other cars. You are likely somewhat safer to be at the side of the road than otherwise, but it obviously is still a dicey situation.

This also brings up another facet. We might all have sympathy for someone that has a car that suffers a mechanical breakdown, doing so while you are driving along on the roadways. Depending upon the nature of the mechanical failure, you might have a difficult time trying to get the car to the side of the road. You might not realize the magnitude of the mechanical failure and end-up stranded in the middle of the roadway. I dare say we all dread such a situation and can be sympathetic to anyone that finds themselves in such a boat.

But, suppose you knew that the person driving the car knew that their car was already having mechanical problems. Thus, they knowingly opted to get onto the roadway and endanger themselves and others. It wasn't as though a lightning bolt from the sky suddenly struck their car, and instead it was something they should have tended toward before it got out-of-control. I think our sympathy would be a lot less.

If that's the case, what kind of sympathy should we have for the person that runs out of fuel? Some would say none. You should have zero sympathy for such people. They were likely forewarned by their car. They drive a car as a privilege granted by the state. They are supposed to be responsible drivers. This includes ensuring that your car has sufficient fuel that it won't run out of fuel. No excuses.

Well, whether or not we should have any sympathy is neither here nor there, herein. The reality is that people do have their cars run out of fuel. It happens.

It can occur by happenstance or it can occur by lack of attention or care. However it happens, once it happens, there's a potential danger for all parties, including those in the car and those anywhere near to the car.

What does this have to do with AI self-driving cars?

At the Cybernetic AI Self-Driving Car Institute, we are developing AI software for self-driving cars. One aspect of interest is what to do about getting low on fuel and how the AI should deal with such a matter.

Allow me to elaborate.

I'd like to first clarify and introduce the notion that there are varying levels of AI self-driving cars. The topmost level is considered Level 5. A Level 5 self-driving car is one that is being driven by the AI and there is no human driver involved. For the design of Level 5 self-driving cars, the auto makers are even removing the gas pedal, brake pedal, and steering wheel, since those are contraptions used by human drivers. The Level 5 self-driving car is not being driven by a human and nor is there an expectation that a human driver will be present in the self-driving car. It's all on the shoulders of the AI to drive the car.

For self-driving cars less than a Level 5, there must be a human driver present in the car. The human driver is currently considered the responsible party for the acts of the car. The AI and the human driver are co-sharing the driving task. In spite of this co-sharing, the human is supposed to remain fully immersed into the driving task and be ready at all times to perform the driving task. I've repeatedly warned about the dangers of this co-sharing arrangement and predicted it will produce many untoward results.

Let's focus herein on the true Level 5 self-driving car. Much of the comments apply to the less than Level 5 self-driving cars too, but the fully autonomous AI self-driving car will receive the most attention in this discussion.

Here's the usual steps involved in the AI driving task:

- Sensor data collection and interpretation
- Sensor fusion
- Virtual world model updating
- AI action planning
- Car controls command issuance

Another key aspect of AI self-driving cars is that they will be driving on our roadways in the midst of human driven cars too. There are some pundits of AI self-driving cars that continually refer to a utopian world in which there are only AI self-driving cars on the public roads. Currently there are about 250+ million conventional cars in the United States alone, and those cars are not going to magically disappear or become true Level 5 AI self-driving cars overnight.

Indeed, the use of human driven cars will last for many years, likely many decades, and the advent of AI self-driving cars will occur while there are still human driven cars on the roads.

This is a crucial point since this means that the AI of self-driving cars needs to be able to contend with not just other AI self-driving cars, but also contend with human driven cars. It is easy to envision a simplistic and rather unrealistic world in which all AI self-driving cars are politely interacting with each other and being civil about roadway interactions. That's not what is going to be happening for the foreseeable future. AI self-driving cars and human driven cars will need to be able to cope with each other.

Returning to the topic of running out of fuel, let's consider how things will differ or be the same in a world that includes AI self-driving cars of the true Level 5 autonomous style.

Presumably, the AI will be able to detect the amount of fuel available in the self-driving car.

I'll for the moment put to the side the circumstances wherein the fuel detection system is faulty. This could happen and I just want to point out that it could happen. This is sometimes a surprising notion to those that believe that AI self-driving cars will somehow have super powers and never fail. What, they ask, you are suggesting that somehow the fuel detection system might not work? Impossible, they say, this is a flawless AI self-driving car that has no imperfections.

I am not sure what makes these people think that AI self-driving cars will never have any issues. I suppose they watch a lot of science fiction movies in which the systems of the future never falter or breakdown. Or, maybe they have fallen for some kind of marketing drivel that showcases AI self-driving cars as shiny vehicles that exude perfection. Anyway, get used to the idea that in the real-world there are going to be AI self-driving cars that suffer mechanical breakdowns, might have recalls, and have the same kind of frailties as conventional cars do.

Hopefully, the AI should be developed such that if the self-driving runs out of fuel, even if the fuel detection system is claiming there is fuel, the AI will have a fallback operation that can then try to get the self-driving car into a minimal risk condition, such as getting over to the side of the road. I realize that the odds of the fuel detection system being so far off that the AI gets caught unawares would seem generally unlikely, but the fact is that it could happen and thus there needs to be a contingency by the AI for this possibility.

We can also ponder whether or not the AI would even be able to realize that the fuel is exhausted. Imagine that the fuel detection system is informing the AI that there is a half tank full of fuel. Meanwhile, the self-driving car starts to sputter and the engine dies. The AI now has a situation of the self-driving car's engine no longer working, but the AI is getting meanwhile no particular indication of why. The fuel volume seems to be just fine, and yet the engine has shut-off.

There is another possibility too involving some other mechanism that has gone awry and won't feed fuel to the engine. Thus, the fuel detection system might be completely correct, and yet the fuel getting from the tank or storage bank to the engine has gone afoul. The point being that there are more than just one means by which the engine might no longer be getting fuel, either due to the lack of fuel or the inability for the fuel to actually reach the engine.

The AI self-driving car should have various sensors and electronic communication going on with the ECU (Engine Control Unit), which then might aid the AI in diagnosing what is going on. This though can be tricky since presumably it has to happen in real-time. Suppose the AI is driving the car on a freeway at 80 miles per hour and all of a sudden, the engine stops running.

There is a need for the AI to try and quickly figure out what has gone amiss. At the same time, no matter what has happened, the AI needs to be taking action about the aspect that the engine is no longer working. You can debate somewhat about how much diagnosis is needed other than knowing that the engine has stopped. On the other hand, if the AI could discern why the engine has stopped, it might open more possibilities of what next action to take.

In short, we assert that the AI needs to have a capability to contend with a situation involving the AI self-driving car running out of fuel.

Some AI developers would say that this is covered by their general approach of having the AI deal with anything that might go amiss with the AI self-driving car. In essence, they claim that running out of fuel is no different than say the self-driving car having a tie rod that breaks. It's all part of the contingency driving aspects routine.

This is not really the case.

If the car is otherwise fully functional, other than the lack of fuel, there is a chance to use the in-motion of the self-driving car to try and cope with the situation. A severe mechanical failure is unlikely to allow the AI the chance of maneuvering the self-driving car and trying to find a safe way to get out of the current difficulty. We eschew this notion that all issues are the same and that there is no need to differentiate the various kind of issues that can arise during a driving journey and the functional capabilities of the self-driving car.

Some AI developers would say that the fuel running out is an edge problem. It is not at the core of what they are aiming to get undertaken with an AI self-driving car. An edge problem is one that is at the corner or edge of what you are otherwise trying to solve. Yes, having the AI self-driving car be able to drive on a roadway and do so without hitting anyone is a core mission. But, not being able to contend with things that can go awry, including the fuel issues, seems like a myopic view and will get true AI self-driving cars into hot water (along with endangering people).

Let's shift gears, so to speak, and now consider the situation of an AI self-driving car that is genuinely getting low on fuel and the AI knows that the self-driving car is indeed getting low on fuel.

What happens then?

Some auto makers and tech firms are skirting the issue right now because they are essentially controlling their AI self-driving cars and so they force the AI to go ahead and do a fill-up. In other words, if you are doing roadway trials with your AI self-driving cars, and you are pampering those AI self-driving cars with a dedicated team of engineers and maintenance personnel, the odds are that you aren't going to allow the AI self-driving car to run out of fuel.

In the wild, once AI self-driving cars are prevalent, and in the hands of consumers, what happens then?

We'll consider two different circumstances, one involving human occupants inside the AI self-driving car when the fuel question arises, and the other is the situation when there is no human inside the AI self-driving car.

Keep in mind that AI self-driving cars will be driving around at times with no human occupants. It could be that the AI self-driving car is trying to get to location where it is going to pick-up humans, or maybe it is being used as a delivery vehicle and so heading to a destination to make the delivery, and so on.

It is predicted that AI self-driving cars will be used extensively for ridesharing purposes. If you owned an Ai self-driving car that was being used for ridesharing, you might have it trolling around as it waits for a potential rider to request a ride. You look out the window of your office and see your AI self-driving car cruising back-and-forth, waiting for someone to request a lift. Why not park the car? There might not be any available parking, plus you might want your AI self-driving car to be the first to respond to a request, and if it is cruising around this might be a better chance than if it is stopped and parked.

If an AI self-driving car has human occupants, and the AI detects that the fuel is getting low, should the AI let the passengers know? If so, should it ask them if it is Ok for the AI to then find a place to get more fuel?

It would be as though you are in an Uber or Lyft ridesharing car of today, and the human driver turned to look at you, while sitting in the backseat as a passenger, and let you know that the car is getting low on fuel. This is informative to you. The driver might then ask if it is OK for the driver to stop at a nearby gas station and fill-up. I'm betting you would be irked by such a question. You are presumably paying to get from expeditiously from point A to point B. Having to stop and have the car get fuel seems like a rather untoward act.

Suppose you refuse the request by the driver. If the knows they cannot get to your destination prior to running out of fuel, it would likely that the human driver would insist that the car must be brought to a fueling station and whether you like it or not, the driver is going to do so. Outrageous, you exclaim! Why didn't the driver beforehand make sure there was sufficient fuel to make it to the desired destination?

This is the same logic that some AI developers tell me about the question of fuel when I ask them what they are doing about fuel levels in their AI self-driving car software.

These AI developers tell me that it won't ever happen that the AI self-driving car will have passengers and be at a low ebb in terms of fuel. The AI is programmed to always be getting more fuel whenever it is otherwise not engaged with a passenger, plus, the moment that a passenger indicates where they want to go, the AI can ascertain whether there is enough fuel and thus refuse to take on the passenger if the fuel would be insufficient.

There are some holes in this logic.

Suppose that an AI self-driving car has taken on-board a passenger that wanted at first to go to the nearby park. The AI calculated the miles involved and figured out how much fuel there is and determines that it can make it to the park. During the driving journey, the passenger says that they need to pick-up their friend that also wants to go to the park, which means a side trip now for the AI self-driving car. Imagine that this then pushes the potential fuel consumption such that the AI self-driving car would not be able to get to the friend and to the park and then to a fueling station.

You've now got a circumstance of a passenger in the AI self-driving car and there is insufficient projected fuel to satisfy the driving journey being requested.

Thus, this fanciful notion that the AI self-driving car would never have a passenger and yet get into a situation of not having enough fuel for a driving journey is shall we say weak.

My point is that there is going to be situations in which the AI self-driving car will have passengers and yet the desire of the passengers might exceed the projected available fuel. The AI should be able to then interact with the passengers and explain the situation. Furthermore, there might need to be an interactive dialogue about what to do. In the case of the side trip to get the friend while on the way to the park, perhaps the AI explains that it needs first to fuel-up if the side trip is to be undertaken, and therefore "negotiates" with the passenger about what to do.

You might say that it is unnecessary to have a dialogue with the passengers and instead the AI can just tell the passengers what is going to happen, regardless of actually what they want or doing any kind of interaction with them.

The AI might simply tell the passenger that wants to go pick-up a friend prior to going to the park that this side trip is not going to happen. Without even particularly saying why, the AI might just emit an indication that the side trip is not being permitted and that's that.

I have a feeling that if AI self-driving cars do that kind of dictatorial driving, people are not going to want to get into an AI self-driving car. People would probably be more willing to deal with a human driver than to have a "robot" that tells them what is going to happen and offers no capacity to try and explain or reason about the driving.

On the other side of this coin, presumably we don't want a human passenger to let the AI self-driving car get into a dicey situation.

Suppose the AI tells the passenger that the side trip will mean that the AI self-driving car would run out of fuel and get stranded. The passenger maybe decides to tell the AI to go ahead and proceed anyway. Is the passenger crazy? We don't know. Maybe the passenger misunderstood the AI and the situation. Maybe it is an emergency of some kind and the risk is worth it to the passenger? Could be various explanations.

Anyway, should the AI allow a passenger, a human, the ability to override what the AI has determined to be the case that the AI self-driving car is likely going to run out of fuel and become stranded, and thus the human is telling the AI that it is Okay to have this happen?

If you are the auto maker or tech firm, you likely would say no, never allow a passenger to override this kind of circumstance. If you are the human passenger and have some kind of rationale for why you want to do this, you are going to perhaps have a different viewpoint on the AI essentially overriding your command to it.

I'll point out that this is not the only boundary of having the AI and a human at potential odds about a driving task. There are other kinds of situations in which the AI is going to want to do one thing, and a human passenger might want to do something else. The industry does not yet have any clear cut means of trying to ascertain when the AI should so proceed versus acquiesce to the wishes of the human. It's a difficult matter, for sure.

Let's next consider the situation when there isn't a passenger in the AI self-driving car. This would seem generally to be an easier circumstance to deal with in terms of the fuel situation.

Suppose the AI self-driving car is trying to deliver a package across town and at first calculated it could make it to the destination without having to re-fuel. Turns out that the AI self-driving car got stuck in traffic due to a car crash that had blocked all lanes, and the fuel of the AI self-driving car got excessively used up due to this unpredicted and unforeseen delay.

I think we might agree that the AI self-driving car should as soon as practical go get fueled-up. This is likely to create an even further delay in delivering the package. Presumably, the AI self-driving car is going to be interacting with some other system or people to let them know about the delay.

Overall, whenever there is not a passenger in the AI self-driving car, I would guess that we would expect the AI to be monitoring the fuel and take care to make sure to go do a re-fuel when needed. There would need to be a sufficient safety margin that whatever kind of unexpected aspect arises, it hopefully can still make it to the refueling.

Suppose though that the AI self-driving car is not able to do so, such as the case of being on the freeway and all lanes of traffic are blocked.

In that case, it could be that the AI has no options to do anything other than sit there on the freeway and use up fuel.

Sure, it can try to minimize the amount being consumed, but let's assume that it ultimately does run out of fuel and has no other means to do anything (it is stuck in traffic and no way to get out).

I know some AI developers that claim this kind of scenario is less odds than a meteor flying down to earth and striking the self-driving car.

I'm not sure they are right about that aspect.

My view is that betting on something never happening and yet that it could happen, seems like a lousy bet.

In an era of AI self-driving cars that are prevalent, presumably the AI self-driving car could use V2V (vehicle to vehicle) electronic communications to let other nearby AI self-driving cars know that it is getting low on fuel.

This might then get the other AI self-driving cars to help out and open a path for the AI self-driving car to get off the road.

I suppose that if the AI self-driving car did run out of fuel on the roadway, perhaps the other nearby AI self-driving cars might come to its aid.

For example, another AI self-driving car might give the stranded one a push to help it get out of traffic. Other AI self-driving cars might block traffic to let this activity take place. Meanwhile, via perhaps V2I (vehicle to infrastructure), an electronic message goes out to a local tow truck that it should come and get the AI self-driving car.

The Kepler space telescope recently ran out of fuel, bringing to an end it's nearly 10-year planet hunting voyage across outer space, having successfully discovered several thousands of exoplanets.

Away it now drifts, looping around the sun and taking on an Earth-trailing orbit. NASA knew that the Kepler would eventually run dry.

Closer here to home, we should be thinking about the situations when an AI self-driving car goes dry, i.e., running out of fuel.

It seems like perhaps a trivial matter in comparison to hunting for new planets, but I assure you that once we have a prevalence of AI self-driving cars, people are going to want to know that those AI self-driving cars can deal with being low on fuel, along with being able to appropriately handle situations of running out of fuel.

I know it seems counter-intuitive that these seemingly "super powered" AI self-driving cars could somehow find themselves stranded, and we would assume that running out of fuel could only happen to distracted or imprudent human drivers, but these kinds of assumptions need to be revisited.

Maybe we'll see an AI self-driving car that one day is pleading for fuel, hey buddy, the AI asks another nearby self-driving car, can you spare a gallon or two (or, perhaps some megawatts)?

CHAPTER 7

DEEP PERSONALIZATION
AND
AI SELF-DRIVING CARS

CHAPTER 7

DEEP PERSONALIZATION

AND

AI SELF-DRIVING CARS

Does your car know you?

In some relatively simple ways, it might.

The other day I rented a car during a trip up to Palo Alto to speak at an industry conference and upon getting into the rental it took me about a solid five to ten minutes to adjust the car to my preferences. I'm guessing you've done the same from time-to-time when getting into a rental car, though hopefully spending less time in doing so (in this instance, I wasn't in a hurry and opted to play around with various features and settings).

The driver's seat was too close to the steering wheel and so I moved it back a few inches. Side mirrors of the car were at an angle and a position that made them unusable for me, so I re-angled them. It was going to be a somewhat lengthy drive in the rental car and so I opted to set the radio station settings for ease of choosing the ones that I like. For the in-car temperature control, I adjusted the settings to fit to my liking. The GPS system was a handy capability and I decided to go ahead and enter my destinations into it.

As an aside, I am continually amazed that when I get a rental car that has GPS that the prior renters often do not erase their destinations and I get a chance to see where other people have gone, which it's not as though I am prying since the prior destinations show-up the moment you enter newly desired destinations. My rule-of-thumb is that I always erase my entered destinations upon returning the car. But, hey, that's just me, always thinking about privacy aspects and cars.

There were some facets of the car that I had to orient myself to and then just live with the settings in whatever manner the auto maker had decided to pre-establish them. Where is the button for locking and unlocking the doors? Where are the cup holders? Where are the emergency flashers if I need them in a hurry? Are the headlights working (one time, I rented a car and had not checked the headlights right away, and later on at night when I needed them, turns out one of the headlights wasn't working)? And so on.

In some respects, I was able to personalize the car to my own preferences, such as the seat settings, the temperature settings, the GPS destinations, etc. There were other aspects that I could not personalize and yet would need to know how to access and use them, even if I could not adjust them to my own preferences. All in all, as mentioned, it took me a few focused minutes to get the car ready for my use.

Another aspect that you typically need to get used to when using a rental car is the nature of the brakes, the accelerator, and the steering wheel. If you are used to driving your own car, you've likely grown accustomed to how much pressure to apply to the brakes in order to bring the car to a halt. Likewise, you've gotten used to the amount of pressure needed to do a rapid acceleration versus a slower acceleration via the accelerator pedal. For the steering wheel of your car, you are likely quite familiar with how much of a turning effort of the steering wheel will cause the wheels of the car to turn.

Since cars typically differ in terms of how sensitive the pedals and steering wheel are, whenever you get into a different car than your own, you often need to figure out what the sensitivity level is like in this "stranger" car that you are going to use. Though I've mentioned this aspect regarding a rental car, you can experience the same sense of sensitivity differences by driving say a friend's car. There's nothing special about getting to know a rental car's sensitivity, it's just a matter of the difference between what you normally drive and the act of driving something else, whether it is a rental car or a friend's car or whatever.

One time, I rented a car that was the same make and model of my car at home. I assumed that the feel of the car would be identical. Turns out that the nature of how the rental agency had maintained the car led to some subtle differences in how the pedals and the steering wheel reacted. It was quite similar to what I was used to driving, but not exactly the same.

If someone drove my car and then drove that rental car, I'd bet that most would not be able to discern a difference. It was only because I was attuned to my own car's aspects that I could discern the rather minor differences. On a macro level, they both seemed to have the same sensitivity. On a micro level, having driven my own car every day for my daily commute, I am "at one" with my own car and how it handles on the road (should I be proud of this fact or does it indicate that I spend too much time on the road?).

I had an embarrassing moment some months ago that when I rented a car at the airport and had several colleagues accompanying me on the trip. We all piled into the rental car and we were in a hurry to go since the flight had been late arriving at the airport. I turned on the engine and knew that we were in a rush, so I skipped my usual "pre-flight" checking of the car and did not do any of my customary preferences settings. Without any delay, I put my foot on the accelerator pedal and gave it a light tap. It was an amount of a tap that on my car at home would have caused the car to barely inch forward.

But, yikes! Unfortunately, this rental was about as sensitive as any car that I've ever driven. A light tap to the accelerator caused the car to gun forward and my passengers were thrown back into their seats. Hey, rocket man, they exclaimed, take it easy and let's aim to get to the event in one piece.

It took me a few minutes of driving throughout the airport area to gradually get used to the sensitivity levels of the accelerator, the sensitivity of the brakes, and the sensitivity of the steering wheel. Usually, if they any of them requires a lighter touch than my car, they all then require a lighter touch. Similarly, if any of them require a heavier touch, they all require a heavier touch. This particular rental car was kind of weird in that the accelerator responded to the lightest touch, while the brakes required putting ten thousand pounds of foot pressure to get the car to slow down, and the steering wheel had an odd aspect that if you turned it clockwise it was easy but if you turned it counter-clockwise it was a battle.

During that trip, one of my colleagues was quite vocal about her preferences of driving style. I tend to be a more conservative driver. I generally abide by the speed limits and like to come to a full stop at stop signs. This was causing my colleague to go berserk. When I rolled up to a stop sign, after having come to full stops several times previously, she said loudly that lead foot was going to once again come to a complete stop and make us all wait for heaven knows why. She was vocal throughout the visit and though I offered to let her drive, she continued to insist that she was Okay with my driving and I should just keep going.

After we had gone to the event, we had time to drive around and look at the local touristy types of things, like driving to see some notable statues, bridges, fancy homes, and the rest. One of my colleagues was especially interested in old-time movie theatres and knew all about them. I drove somewhat in a meandering fashion throughout the town and whenever we saw a classic movie theatre, I'd pull over, so we could admire it for a few minutes.

In a sense, I was trying to personalize the driving experience for my colleagues. This included driving in a manner or style that would be comfortable for them. It also included driving to places that they might like. In addition, it involved driving slowly and possibly idling to see something, while in other instances driving quickly whenever there wasn't anything notable for them see.

I'm sure you've done the same.

Perhaps when you have visitors come to your town or city, you drive them around. In doing so, where you drive and how you drive might differ substantially from how you and drive and where you drive on a normal basis. Here in Southern California, we have the world-famous Disneyland, and any visitor that comes to see me is usually eager to go see the theme park. Meanwhile, having seen Disneyland a million times already, I've gotten used to it and barely even notice it when driving past Mickey Mouse and his pals (well, Ok, I admit that I get a big smile and instantly get that goosebumps feeling).

Overall, your car at home has likely been personalized by the means of how you have set the seats, the temperature controls, the radio stations, and the like. You also presumably know by heart the sensitivity of the driving controls, knowing how much pressure to apply to the pedals and to the steering wheel to produce a smooth motion of the car (or, whatever motion of the car that you prefer). You have gotten used to the local areas that you drive, and so you know shortcuts and places to go, along with areas to avoid.

What does this have to do with AI self-driving cars?

At the Cybernetic AI Self-Driving Car Institute, we are developing AI software for self-driving cars. One aspect for AI self-driving cars is the potential for them to provide deep personalization.

Personalization refers to how well your AI self-driving car knows you and your preferences, along with the AI trying to fulfill those preferences as best it can.

There is a range of personalization.

In some cases, the personalization might be relatively shallow and not provide much of any personizing to you. In other circumstances, the personalization can be deep in that the AI has been able to deeply get to know you and your preferences and abides by those preferences when feasible.

For most of the auto makers and tech firms that are developing AI self-driving cars, they consider this notion of personalization to be an edge problem. An edge problem is a corner case or considered at the edge of the core aspects that you are trying to solve. Right now, the auto makers and tech firms are focused on getting an AI self-driving car to do the normal things that you would want a self-driving car to do, such as driving down the road and not hitting anything along the way.

I won't get into a debate herein about whether or not personalization is an edge problem per se.

Admittedly, an AI self-driving car that can drive according to your personal preferences is certainly not as high a priority as getting the AI self-driving car to drive the car overall. It doesn't do much good to have personalization if the AI cannot even drive the car properly and appropriately. Nonetheless, I'd assert that humans will generally want to have personalization when being an occupant in an AI self-driving car and that ultimately it is something that will be worthwhile to have included into the AI self-driving car capabilities.

I'd like to first clarify and introduce the notion that there are varying levels of AI self-driving cars. The topmost level is considered Level 5. A Level 5 self-driving car is one that is being driven by the AI and there is no human driver involved. For the design of Level 5 self-driving cars, the auto makers are even removing the gas pedal, brake pedal, and steering wheel, since those are contraptions used by human drivers. The Level 5 self-driving car is not being driven by a human and nor is there an expectation that a human driver will be present in the self-driving car. It's all on the shoulders of the AI to drive the car.

For self-driving cars less than a Level 5, there must be a human driver present in the car. The human driver is currently considered the responsible party for the acts of the car. The AI and the human driver are co-sharing the driving task. In spite of this co-sharing, the human is supposed to remain fully immersed into the driving task and be ready at all times to perform the driving task. I've repeatedly warned about the dangers of this co-sharing arrangement and predicted it will produce many untoward results.

Let's focus herein on the true Level 5 self-driving car. Much of the comments apply to the less than Level 5 self-driving cars too, but the fully autonomous AI self-driving car will receive the most attention in this discussion.

Here's the usual steps involved in the AI driving task:
- Sensor data collection and interpretation
- Sensor fusion
- Virtual world model updating
- AI action planning
- Car controls command issuance

Another key aspect of AI self-driving cars is that they will be driving on our roadways in the midst of human driven cars too. There are some pundits of AI self-driving cars that continually refer to a utopian world in which there are only AI self-driving cars on the public roads. Currently there are about 250+ million conventional cars in the United States alone, and those cars are not going to magically disappear or become true Level 5 AI self-driving cars overnight.

Indeed, the use of human driven cars will last for many years, likely many decades, and the advent of AI self-driving cars will occur while there are still human driven cars on the roads. This is a crucial point since this means that the AI of self-driving cars needs to be able to contend with not just other AI self-driving cars, but also contend with human driven cars. It is easy to envision a simplistic and rather unrealistic world in which all AI self-driving cars are politely interacting with each other and being civil about roadway interactions. That's not what is going to be happening for the foreseeable future. AI self-driving cars and human driven cars will need to be able to cope with each other. Period.

Returning to the topic of personalization, I categorize the potential levels of personalization of AI self-driving cars into these classes:

- No personalization

- Shallow personalization

- Substantive personalization

- Deep personalization

The lowest level, consisting of no personalization, gets assigned when there is no particular attempt at personalization and that by any reasonable judgment we would likely agree that there is nothing really built into the AI self-driving car system for personalization purposes.

The shallow form of personalization consists of attempts at personalization that seem rather token and flimsy. Substantive personalization would be personalization that is genuine in nature and spirit, providing a somewhat convincing set of personalization aspects. The topmost category is the deep personalization mode. For deep personalization, the AI needs to have gone all out to provide a wide and in-depth set of well-coordinated and aligned personalization capabilities

At this time, most of the auto makers and tech firms are by default aiming at no personalization (or, at best, shallow personalization), meaning that they are not devoting key resources or attention toward personalizing their AI self-driving car technology due to it being considered an edge problem. They have in mind to eventually get around to adding personalization, but it is not in the cards for years to come, until they've go squared away with the other self-driving car foundational elements.

One AI developer told me that they are indeed going to have personalization that launches with the release of their true AI self-driving car, and when I asked if he could describe in general terms what the personalization feature would be able to do, he explained proudly that the AI would be able to say the name of the owner of the car. For example, suppose Samantha has bought an AI self-driving car and whenever she uses the self-driving car it would greet her by saying "Hello, Samantha" and while in the car and on a journey the system would say things like "Samantha, only fifteen minutes until reaching your destination."

I could not believe that this AI developer was serious about claiming that this was personalization. The mere act of repeating aloud the name of the owner is about as lame as you could get in the personalization categorization scheme. The AI would not know anything about the person and would be simply uttering the sound "Samantha" as inserted into templates of Natural Language Process (NLP) that was being designed for interaction with the car occupants.

I asked what happens if someone else borrows Samantha self-driving car, what would the AI say to that person? He indicated that when the person gets into the self-driving car, the AI would ask if the person was Samantha, and if the person said they are not Samantha, it then would not do any "personalization" and thus would not say the word "Samantha" anymore.

I felt like I had landed back into the 1980s of computer technology. I could not decide whether to classify this as "No Personalization" or perhaps to satisfy the AI developer's claim that there was "some" amount of personalization that I might add a new sub-classification known as Ultra Shallow Personalization.

I pointed out to him that there are some potential adverse consequences related to his so-called personalization. One being that if the human occupant believes that the AI knows more than it does, the human might make undue assumptions about what the AI is capable of doing, and the human might get themselves and the AI self-driving car into some untoward spots. By parroting the name of the person, it could suggest that the AI knows a lot more than it does. Anthropomorphizing automated systems can lead to false understandings by humans about what the technology can and cannot do.

I also was a bit taken aback that the AI developer and their firm had apparently not considered some relatively easy ways to augment this notion of using the person's name (assuming that's the route they wanted to go with). Allow me to elaborate.

The AI self-driving car has sensors such as cameras, some pointing outward and some pointing inward, and it would be relatively straightforward to have the cameras do a facial recognition match to anyone that approaches or that gets into the self-driving car. From that aspect alone, at least the AI could likely ascertain whom Samantha was (after initially establishing her in the system), and automatically henceforth be able to say her name when appropriate (and not say her name when appropriate).

I'm not suggesting that the AI would know anything about Samantha per se, but at least it could do the legwork of using facial recognition to likely ascertain when she is near to the self-driving or within the self-driving car. This is something that they could do with today's technology and would not require any significant leap or advancement to craft.

Consider too what else could take place if the facial recognition was adopted.

As an example, this would allow potentially for the AI to unlock the car doors to let Samantha into the self-driving car.

If she seemed to be walking toward the self-driving car and upon the facial recognition recognizing her, the AI could unlock the car doors based on the assumption that she was likely to want to get into the self-driving car. If she was walking nearby but not toward the self-driving car, the AI might be "smart" enough to realize not to yet unlock the car doors and wait until the moment seemed more opportune for the action (otherwise, it might become frustrating to the person that each time they neared their self-driving car it kept unlocking itself, which maybe was not what the person actually desired to have happen all of the time).

There is also the possibility of allowing Samantha to provide commands to the AI self-driving car while being outside of the AI self-driving car.

Suppose that Samantha wanted the AI system to go over to the grocery store and pick-up her awaiting groceries that had been picked-and-packed by a clerk at the store based on an order she placed via her smartphone. She might walk over to her AI self-driving car and tell it to go to the grocery store, wait there and pick-up the groceries, and then come back to the house.

If Samantha had done this kind of action previously, the AI could potentially anticipate such a command and even ask Samantha as to whether that's what she wanted to have undertaken.

This then starts us down the path of understanding true personalization in terms of having the AI be able to identify patterns of behavior that can then be "learned" about over time by the AI and then reflected in what the AI does related to the actions of the self-driving car.

One feature that some cars have today consists of allowing you to set the driver's seat in terms of its position forward, its tilt, its heating or cooling pad capability, and so on. Once you've set things to how you like the seat, you can have the seat configuration "memorized" (actually, just stored in memory of the computer), and later on if someone messes with your seat settings you can have it return to your personalized settings.

The AI of the self-driving car can do that same kind of "memorization" and yet go even further by being able to consider the context of the circumstances. Context can be crucial to have personalization that makes sense versus personalization that is "dumb" and simply taking place on a rote basis. Recall that I mentioned that when Samantha approaches the self-driving car that it would not always necessarily immediately unlock the door. Always unlocking the door would be a simplistic rote kind of approach, akin to the seat configuration memorization of today's cars.

I had earlier told you about the time that I was driving several of my colleagues and one of them got mildly upset that I was such a conservative driver. Yes, I am normally a law abiding and sensibly cautious driver. That does not mean though that I always drive that way. If there's an emergency and I need to quickly get to the nearest hospital, I assure you that I am going to put the metal to the floor. The context determines to some extent the nature of the driving style that I would choose to use.

Let's use another example to consider the importance of context in personalization.

Suppose that Samantha has traveled in her AI self-driving car many times. She has especially used it to get to work, along with getting her to the gym after work.

The AI, if well prepared with personalization capabilities, would be silently noting the use of the self-driving car by Samantha. Times of day and days of the week that the self-driving car takes her to work and when she goes to the gym. Various driving paths to get to those destinations and the variants of traffic conditions would be tracked. There are perhaps some instances wherein she was in a rush and asked the AI to try and as quickly as feasible get to work. And so on.

Within the AI self-driving car, there isn't a driver's seat, but there are other seats for the occupants. Perhaps Samantha prefers to sit in the seat on the right-side of the self-driving car and prefers in the morning commute that it be swiveled inward so that she can catch-up on her reading and preparations for the work day ahead. After work, she prefers to have the seat swiveled to readily look out the window of the self-driving car and watch the scenery as she heads home or to the gym. And so on.

These are all facets of her behavior related to the AI self-driving car. The AI can keep track of these aspects, doing so not as standalone matters but within the overall context of how they arise. By analyzing the collected data, the AI can potentially spot trends and preferences of Samantha, making use of Machine Learning (ML) and then make use of those gleaned insights to personalize the behavior of the AI self-driving car to suit her needs.

I realize that this gathering of data about Samantha and her behavior, along with analyzing it, raises questions about privacy aspects. The auto makers and tech firms will be facing some tough issues about the degree to which the AI should and should not do this kind of tracking. Furthermore, will Samantha know that she is being tracked? Wil she be able to switch on or off the tracking?

Another question involves the storage of this tracking data. We might assume that the data about Samantha is stored locally in the on-board AI system of the self-driving car. But, there is also the use of OTA (Over-the-Air) features for AI self-driving cars, allowing for data to be shared up into the cloud of the auto maker or tech firm, and also for the auto maker or tech firm to pump updates and patches down into the AI self-driving car.

The tracking data about Samantha and her use of the AI self-driving car could readily be pushed up into the cloud of the auto maker or tech firm. You might wonder why they would do so.

Presumably, the data could be used for aspects such as fine tuning the AI system based on the kinds of behavior that people are exhibiting inside the AI self-driving cars of that auto maker or tech firm. In essence, they could potentially anonymize the data and try to find patterns collectively across all those that are using their brand of AI self-driving cars. This is known as fleet learning.

Another potential use might involve being able to readily personalize other AI self-driving cars of the auto maker or tech firm whenever Samantha uses a different such self-driving car.

Suppose she is at work and has let a friend make use of her AI self-driving car. She heads out to the curb and hails an AI self-driving car that is driving along as a ridesharing service. She gets into the AI self-driving car. Based on facial recognition, it might have grabbed Samantha's tracking history from the cloud of the auto maker or tech firm, and now this AI self-driving car can act as personalized as the one that she owns.

These are all happy face scenarios about how the tracking of Samantha's behavior is being used.

Of course, there are sad face scenarios too. An auto maker or tech firm could potentially mine the data and then opt to sell the data to third parties or use it to beam advertisements to her. I've implied this is a sad face scenario, though it could also be a happy face scenario if Samantha had indicated explicitly that she wanted this data usage to occur and perhaps she got compensated somehow for allowing her tracking data to be used in this manner.

The deep personalization would not solely be focused on the "owner" of an AI self-driving car. Imagine that Samantha has a husband, two children, and a dog. The AI would have the capability to track each of their respective behaviors, including the dog (keep in mind that AI self-driving cars will likely be used to transport people's pets, in addition to transporting humans). Also, perhaps Samantha let's her friends and co-workers use her AI self-driving car, which they too could then be tracked in terms of their behavior related to the use of the self-driving car, allowing for personalization for them too.

When you consider that AI self-driving cars will most likely be used as ridesharing vehicles, there is going to be ample opportunity to track the behavior of humans in terms of their use of AI self-driving cars. Samantha might decide to earn some extra bucks off her AI self-driving car by allowing it to roam during the work hours and serve as a ridesharing service. The people using her AI self-driving car are then potentially encompassed by the personalization capability.

When I've spoken about the use of deep personalization at AI self-driving car industry events, some inquisitive person invariably asks the rather pointed question of whether this might be creepy. Yes, I certainly agree there is a kind of "creepiness" factor to this. Though, you can say the same about any kind of personalization service (especially the "deeper" or more highly personalized it gets, rather than shallowly personalized versions which are easier to shrug off).

Creepiness can be in the eye of the beholder, which perhaps this next personal anecdote might so reveal.

Years ago, I was doing some consulting work for an east coast client and the client arranged for my hotel stay nearby their headquarters. I arrived at the hotel around dinner time and got out of a cab that had brought me from the airport to the curb directly in front of the hotel. The bellman grabbed my bags from the trunk of the cab and told me to proceed to the front desk, not needing to wait for him.

When I got up to the front desk, the check-in clerk greeted me by using my name. I was puzzled that the clerk would know my name, since I hadn't yet stated what my name was. I asked how she knew my name. She explained that the bellman had looked at my luggage tags and then had radioed to her that I was coming up to the front desk to check-in.

Clever or creepy?

I then went to my room with my bags. I left them sitting in the room, still closed up, and was in a hurry to meet with the client for dinner. I figured that after dinner and when I got back to the hotel room, I'd then have time to open my bags and remove the items that I wanted to store in the room. Meanwhile, I had been reading a book and placed it onto the bed, face down, at the place where I was reading, so that I'd know what page I was on when I returned to the room.

After a very pleasant dinner with some fine wine, I managed to ultimately get back to my hotel room. When I came into the room, the lights were dimmed, and I could see that the bed had been turned down. I realized that the nighttime service that you see at some hotels had taken place. I'm not normally very keen on someone coming into my room, but anyway it is customary in many hotels and didn't seem especially unusual.

I then noticed that the drawers in the room were slightly ajar and the clothes closet was slightly open. This got my curiosity going. When I opened one of the ajar drawers, I saw that my various underclothing and socks had been neatly placed into the hotel room drawers. When I looked in the closet, my shirts and coats had been neatly placed on hangers.

They had opened my bags and opted to put my belongings throughout the room for me.

As a topper, they had put some bedtime slippers and a robe on the bed, which I've seen before, but the kicker was that my book was sitting on the nightstand next to the bed. At first, I was a bit irked because I had purposely placed the book open faced on the bed so that I would not lose the spot in the book that I was last reading. When I went to get my book, I realized they had put a bookmark at that point in the book, and it was even one of those bookmarks that comes with its own little nighttime reading light.

Clever or creepy?

In my own case, I thought it was rather creepy. They had not asked me whether I wanted to have this done. Even if it was considered their standard operating procedure (SOP), it seemed to me that they should have first established that I wanted this done. They could then have placed something into my hotel records that indicated I either wanted this to be done or not, and on subsequent visits have used that "memorization" to guide their actions.

When I told others about what had happened, some of my friends thought it was a tremendous service and I was being thin skinned about it. Others agreed with me that it was rather extraordinary and actually quite odd. One even said to me that he would have started looking around to see if the room had any cameras or audio listening devices. He figured that if they went this far on my bags, who knows what other surprises they had in store. It was a good point and I admit that I was somewhat on edge the rest of the time there (you might find of interest that on later visits to this client, I opted to book at a different hotel, thus, their form of personalization had the opposite effect of what they presumably intended).

Conclusion

Deep personalization for AI self-driving cars carries with it the rather incredible possibility of leveraging the AI of the self-driving car to provide a fully personalized experience for the occupants. It is even quite conceivable that the brands of AI self-driving cars will perhaps be differentiated from each other by which ones do no personalization versus those that do, and ultimately a differentiation between those that do some amount of personalization versus those that do so deeply.

To achieve personalization, deep or otherwise, the AI has to have this capability explicitly built into it. There is no magic wand that somehow allows the AI to just miraculously do personalization.

For those auto makers and tech firms that aren't yet focusing on the personalization capability, at least they should be providing a spot to plug-in such a component, being ready to add the feature when it is built and tested. This capability could then be loaded into the AI self-driving car via the OTA and thus not necessarily need to have the AI self-driving car go to a dealer or auto shop to gain the personalization capabilities once they are ready for adoption.

The thorny topic of how to best provide the deep personalization is something that will need to be gradually figured out, gauging the public's reaction and also potentially whether any regulations arise around the facets of it. Let's aim to have the deep personalization be something in hot demand and that will inspire people to accept and use AI self-driving cars. We don't want to have creepiness that causes people to be worried about a Big Brother kind of AI watching their every move.

Deep personalization for the good of mankind needs to be the mantra of AI developers for self-driving cars.

CHAPTER 8
REFRAMING THE LEVELS
OF
AI SELF-DRIVING CARS

Lance B. Eliot

CHAPTER 8

REFRAMING THE LEVELS
OF
AI SELF-DRIVING CARS

The emergence of Level 3 self-driving cars is going to endanger the advent of AI self-driving cars. There, I've said it. Not many have been willing to stand-up and make that statement. There is though a growing contingent of industry insiders that are gradually voicing their concerns on this matter. It is a serious matter worthy of explicit attention and discussion.

At the recently undertaken World Safety Summit on Autonomous Technology there was some concerted deliberations and debate about Level 3. Indeed, a white paper by the company Velodyne LiDAR entitled "A Safety-First Approach to Developing and Marketing Driver Assistance Methodology" was distributed at the Summit and it essentially asserted that it might be timely to consider reframing Level 2 and Level 3 of AI self-driving cars. I agree with their assertion.

I'd like to walk you through the nature of the Level 3 perils and get you involved in this crucial discussion.

This topic also pertains to the nature of AI and how we can best express the spectrum or range of AI capabilities for a particular task.

In the case of AI self-driving cars, there is the SAE (Society for Automotive Engineers) standard known as the "Surface Vehicle Recommended Practice," numbered as document J3016, which provides a taxonomy and vocabulary for expressing the AI capabilities of self-driving cars. It is a cornerstone to the field of AI self-driving cars. Without it, we'd all have a difficult time even discussing AI self-driving cars, since we wouldn't all have a common set of definitions and meanings.

Some might say that a rose is a rose by any other name, but I assure you that if I am calling a rose an apple, and you are calling a rose an orange, we would have a lot of confusion whenever you said the world "apple" and I said the word "orange." Thus, we do need a foundation of the words we will use to describe something, and in the case of AI capabilities it is essential since we otherwise cannot readily compare one AI system to another one.

Cars like the Tesla outfitted with AutoPilot are currently considered Level 2 on the SAE standard. Some argue that it is more like 2.5 rather than 2, but I point out that it is nonsensical to refer to fractions since the standard definitions are a range of 0 to 5, integers, and there is no such thing as fractionally meeting the defined levels.

Furthermore, it confuses discussions since by making up a fractional level it implies there is such a thing, and it also opens a can-of-worms as to what exactly at 2.5 level consists of. This might also cause a slippery slope toward people opting to refer to 2.8 or 3.1 or any other kind of made-up fractional amount. So, I urge those of you wanting to refer to levels by fractions to stop doing so, thanks (or, I suppose, work toward changing the SAE standard to include fractional levels, if you feel such a compelling need to have them!).

Many of the auto makers and tech firms are now aiming toward Level 3 self-driving cars. I'll explain in a moment the nature of a Level 3 in terms of the AI self-driving capabilities. This will take us from a world somewhat getting accustomed to a Level 2 (though only a small fraction of society has experienced it) and shift us into a world of Level 3 self-driving cars. It is going to happen somewhat incrementally, meaning that we'll see auto maker after auto maker bringing their Level 3 self-driving cars to the marketplace. This will also lead to an uptick in consumer expectations about what AI self-driving cars can and cannot do.

Once I've explained the Level 3 perils, you'll perhaps then see the logic that some are worried about the Level 3 being akin to the famous story about the frog in the boiling water. If you don't know the frog story, here it is. In the mid-1800's, some scientists reported that if you put a frog into water and very slowly brought the water to a boil, the frog would not try to jump out of the pot of water and instead would die in the boiling water, presumably not having been able to have anticipated that their death would arise over the course of the upticks in temperature. On the other hand, if you tossed a frog into a pot of boiling water, it would immediately react and try to escape.

The overall lesson to be learned is that we can sometimes gradually get used to something, but for which in the end it is disastrous for us, and yet had we gotten abruptly tossed into it toward the end we would have been able to readily see that the end was near.

These frog experiments have since been pretty much debunked, but in any case, it is a handy metaphor for referring to situations in which something slowly happens and you kind of gradually fall into it (even if it isn't really true about frogs in boiling water!). I often refer instead to such situations as a form of quicksand. Once you get yourself into quicksand, it at first doesn't seem overly alarming if you are unaware of what quicksand can do. You figure at first that you can somehow readily work your way out of it. Unfortunately, often it is not the case that you can, and so instead you gradually and inexorably get pulled under.

So, Level 3 for some industry participants and observers is the frog in the progressively boiling water or if you like instead it is the quicksand that will sink us all — not only in Level 3, but perhaps sink the rest of AI self-driving car adoption too (or at least substantially undermine the acceptance of AI self-driving cars).

I would dare say that there could be an even larger spillover into AI systems of many varieties, in the sense that if the public and regulators become disturbed at what happens in the AI aspects of AI self-driving cars, it could readily carryover into AI systems of other kinds.

There are some auto makers and tech firms that are purposely avoiding the Level 3 and instead jumping straight to Level 4 and Level 5. One reason to skip past Level 3 entails the perils that I'll be describing for you momentarily. Another reason that some are jumping past Level 3 is that they simply believe in what some would call truly autonomous cars and they don't want to mess around with or get mired in anything less than something that is truly autonomous. You could say they are doing so on an overall technological and business philosophical basis about the nature of AI self-driving cars.

At the Cybernetic AI Self-Driving Car Institute, we are developing AI software for self-driving cars. We too are desirous of achieving truly autonomous cars. This aim at truly autonomous cars is more of a moonshot than might seem to be the case on the surface (for those of you that aren't directly involved in developing such AI).

Allow me to elaborate about the levels of AI self-driving cars.

The topmost level is considered Level 5. A Level 5 self-driving car is one that is being driven by the AI and there is no human driver involved. For the design of Level 5 self-driving cars, the auto makers are even removing the gas pedal, brake pedal, and steering wheel, since those are contraptions used by human drivers. The Level 5 self-driving car is not being driven by a human and nor is there an expectation that a human driver will be present in the self-driving car. It's all on the shoulders of the AI to drive the car.

For a Level 4, the self-driving car is allowed to have various self-imposed restrictions on when the AI can and cannot perform the self-driving of the car. An auto maker might define an Operational Design Domain (ODD), such that their particular version of a Level 4 self-driving car can do the self-driving when the roads are dry and there is no snow, but if a driving situation arises outside of the defined ODD (let's say it begins to snow), the AI is supposed to check to see if a driver is present that might want to intervene, and if not the AI then will perform a fallback effort and put the car into a minimal risk condition (such as pulling off to the side of the road).

For self-driving cars less than a Level 4, there must be a human driver present in the car. The human driver is currently considered the responsible party for the acts of the car. The AI and the human driver are co-sharing the driving task. In spite of this co-sharing, the human is supposed to remain fully immersed into the driving task and be ready at all times to perform the driving task. I've repeatedly warned about the dangers of this co-sharing arrangement and predicted it will produce many untoward results.

Here's the usual steps involved in the AI driving task:

- Sensor data collection and interpretation

- Sensor fusion

- Virtual world model updating

- AI action planning

- Car controls command issuance

Another key aspect of AI self-driving cars is that they will be driving on our roadways in the midst of human driven cars too. There are some pundits of AI self-driving cars that continually refer to a utopian world in which there are only AI self-driving cars on the public roads.

Currently there are about 250+ million conventional cars in the United States alone, and those cars are not going to magically disappear or become true Level 5 AI self-driving cars overnight.

Indeed, the use of human driven cars will last for many years, likely many decades, and the advent of AI self-driving cars will occur while there are still human driven cars on the roads.

This is a crucial point since this means that the AI of self-driving cars needs to be able to contend with not just other AI self-driving cars, but also contend with human driven cars. It is easy to envision a simplistic and rather unrealistic world in which all AI self-driving cars are politely interacting with each other and being civil about roadway interactions.

That's not what is going to be happening for the foreseeable future. AI self-driving cars and human driven cars will need to be able to cope with each other.

Human Attention to the Driving Task

As mentioned, the levels less than a Level 4 have as a clear-cut requirement that a human driver must be present in the AI self-driving.

The human driver must be licensed to drive. The human driver must be ready at all times to take over the driving task. The human driver must be attentive to the driving of car by the AI system, since otherwise the human driver might be aloof of the driving situation and therefore be unable to readily and immediately takeover the driving task.

You've likely seen YouTube videos of human drivers that are sitting in the driver's seat of an AI self-driving car at a Level 2, and those human drivers are not watching the road.

Instead, you see them trying to type on their smartphone, or they are reading a book in their lap, or they are looking at the backseat of the car where their cute baby is cooing at them, etc. There are many such videos, including ones that show the human driver putting their hands outside the driver's side window to convince us that they are not driving the car.

Imagine that in the moment that their arms are waving outside of the car that all of a sudden, a truck veers unexpectedly into the path of the self-driving car and the AI urgently requests that the human driver take over the driving task.

Do you think that the human driver would be able to pull their arms back into the car, put their hands onto the steering wheel, and then sufficiently hit the brakes or maneuver the car to safety, doing so in the likely split seconds before a fatal crash? I think not.

I say that the videos help illustrate a very crucial aspect about human behavior.

If humans believe that an AI system is going to be able to undertake a certain task, it is going to be the case that the humans will get lax in performing the task and become complacent or ill-prepared to undertake the task when the time comes for them to do so.

Take a look at Figure 1 (see next page).

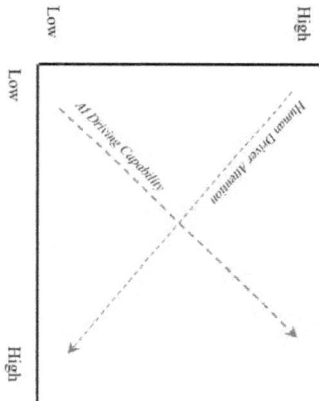

Eliot Framework: Human Driver Attention vs. AI Driving Capability

Figure 1

Copyright © 2018, Dr. Lance B. Eliot, Cybernetic Self-Driving Car Institute.

In the diagram, there is a line that starts at the High position of the vertical axis and then proceeds downward across the diagram toward the rightmost edge.

This line represents the attention level of the human driver. There is a second line on the diagram, which starts at the Low position of the vertical axis at the leftmost edge, and this line makes its way upwards across the diagram.9936 This is a line representing the AI driving capabilities of a self-driving car.

What this diagram is suggesting consists of the notion that as the AI driving capabilities rise, the human attention to the driving task decreases. At the leftmost edge of the diagram, the human driver attention is quite high, which makes sense since the AI driving capabilities are quite low. The human realizes that they cannot rely upon the AI, and so they drive the car as though they are the driver of the car (which, they are). At the rightmost edge of the diagram, the AI driving capabilities are quite high, and as a result the human driving attention is low since the human is fully expecting the AI to handle the driving task.

For those of you that are mathematicians or statisticians, I realize you might want to argue about the diagram in terms of whether there is a direct inverse correlation between these two lines. Also, you might want to argue that the lines shouldn't be portrayed as straight arrows but perhaps have some other more fluid shape to them. Sure, I'll go along with those aspects and want to emphasize that the diagram is meant to overall showcase the nature of the phenomena, and not necessarily be an exact portrayal of it.

Now, please take a look at Figure 2.

Figure 2

Eliot Framework: Human Driver Attention vs. AI Driving Capability

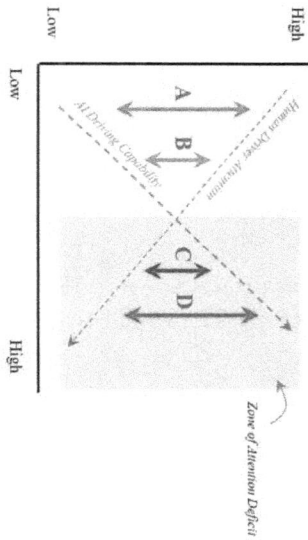

Copyright © 2018, Dr. Lance B. Eliot, Cybernetic Self-Driving Car Institute.

In this diagram, I have highlighted the gaps between the two lines.

The lines eventually cross each other. For the portion that exists to the left of the crossover point, we have a "positive" gap between the level of the human driving attention and the level of the AI driving capability. Once we've reached the crossover point, there is essentially a "negative" gap to the right of the crossover, meaning that the human attention level has now fallen below the level of the AI capabilities (this is a region of the graph that I refer to as the zone of attention deficit).

I've labeled four examples, shown as A, B, C, and D on the diagram, indicating gaps between the human driving attention level and the AI driving capability. Let's consider each of those gap instances.

Example A

For the example labeled as A, we have a relatively large gap of a "positive" nature that shows the human to be at a relatively high level of attention, and this corresponds to a relatively low level of AI driving capability. You can consider this as a good situation generally since it implies the human is aware of the need to be attentive to the driving task when the AI is not up-to-snuff.

Example B

For the example labeled as B, we have a smaller gap of a "positive" nature that shows the human level of attention is getting less and less, and meanwhile the AI driving capability is becoming more and more. This could be Okay if the AI driving capability is sufficiently up-to-snuff and if the human driver can safely rely upon the AI to do the driving task.

Example C

For the example labeled as C, we now have the human attention to the driving task that has dropped below the level of the AI driving capability. The human is now allowing their attention to lapse as they believe that the AI driving capability is able to handle the driving task.

Example D

For the example labeled as D, there is now a large gap between the human level of attention to the driving task, which has dropped quite low, and meanwhile the AI driving capability is indicated as quite high. This is a situation of potential heightened risk since the human is assuming that the AI will indeed be able to handle the driving task.

One aspect to keep in mind about this diagram is that there is the actual AI capability of driving versus the human perceived AI-capability of driving.

The human might or might not accurately understand what the AI self-driving capability actually consists of.

There are many that are concerned that there are mixed messages being portrayed to human drivers about the nature of the AI capabilities in self-driving cars. As such, the human is potentially basing their own attentive levels on what might be a false understanding or combobulated interpretation of what the AI self-driving can actually do.

Let's see how this indication of human attention levels and AI driving capabilities can apply to driving situations.

During a driving journey, you are likely to have times that the driving is somewhat monotonous. Whenever I drive from Southern California up to Silicon Valley, much of the drive consists of an open highway or freeway that usually has little traffic. The driving becomes a rather mindless task and consists of being on the watch for outliers.

Take a look at Figure 3 (see next page).

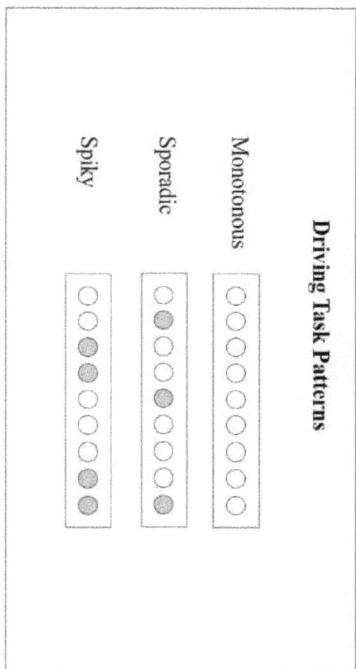

Eliot Framework: Human Driver Attention vs. AI Driving Capability

Figure 3

You can have times during a driving journey that there is sporadic activity. While on the freeway during my morning commute, the traffic is snarled and so much of it takes place on a stop-and-go basis. You come to a stop, you wait, and traffic continues. This repeats itself. In that sense, you could say it is sporadic. It happens with a certain amount of regularity.

There are also situations wherein driving consists of sudden actions. I'll refer to this as spiky driving conditions. Suppose I'm in stop-and-go traffic, and all of a sudden, the traffic opens up and there is a brief period that allows you to quickly accelerate and cover some ground. But, then it comes to a somewhat abrupt halt and you find yourself once again in the classic stop-and-go formation.

Next, take a look at Figure 4 (see next page).

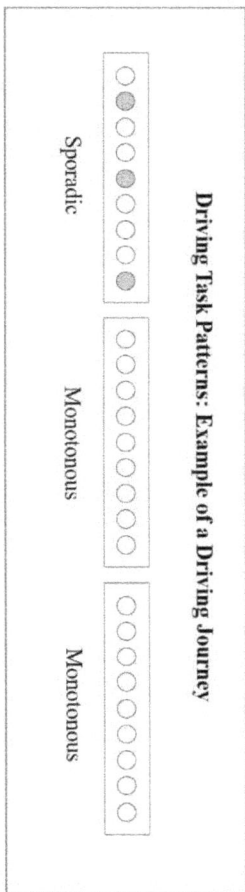

Eliot Framework: Human Driver Attention vs. AI Driving Capability

Figure 4

Driving Task Patterns: Example of a Driving Journey

Sporadic

Monotonous

Monotonous

Copyright © 2018, Dr. Lance B. Eliot, Cybernetic Self-Driving Car Institute.

A driving journey can consist of many segments, each segment being either one of the possibilities of being monotonous, sporadic, or spiky.

While driving home at night after work, I have stretches of my drive that are monotonous, and then find myself in a spiky situation, and then maybe back to being monotonous, and then it perhaps becomes sporadic. And so on.

Shown in the diagram is a driving journey that began as sporadic, and then became monotonous thereafter.

Take a look at Figure 5 (see next page).

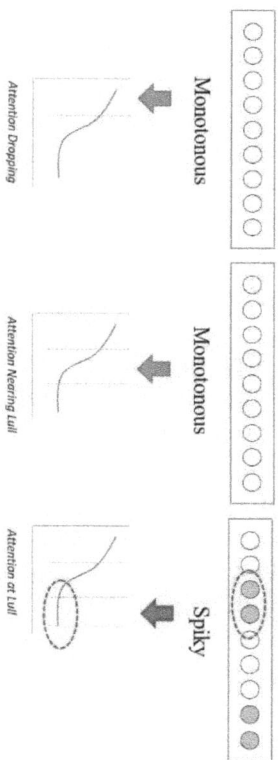

Eliot Framework: Human Driver Attention vs. AI Driving Capability

Figure 5

Driving Task Patterns: Driving Journey and Attention Decay Curve

Human Attention Decay Curve

Monotonous — Attention Dropping

Monotonous — Attention Nearing Lull

Spiky — Attention at Lull

Copyright © 2018, Dr. Lance B. Eliot, Cybernetic Self-Driving Car Institute.

I now want to bring your focus to the human attention that might occur during each of those driving segments.

Depending upon the arrangement or series of driving segments involved, a human driver might become increasingly less attentive to the driving task over time. In other words, perhaps at the start of the driving journey the driver was highly attentive, but then became lulled due to the nature of the driving effort involved.

I am illustrating this without regard for any kind of AI driving capability. This aspect of driving attention applies to situations of conventional car driving, along with the advent of AI self-driving cars.

As indicated in the diagram, the driver attention consists of a decay curve.

In this example, the elongated set of monotonous driving has led to the human attention span dropping. Unfortunately, a spiky instance then arises, and the driver is at low point in their driving attention. You've perhaps experienced this in your own driving. It might be late a night, you are driving for a while, the road is relatively open and uneventful. All of a sudden, a drunk driver pulls onto the roadway. Your attention has been dulled and your reaction time is sluggish. Your ability to react quickly and aptly might be a lot less so than if the drunk driver had pulled onto the road when you first began your driving trip and were more alert and aware.

This is a bad combination in that the driver attention is very low and yet the need to react and respond is very immediate.

The Level 3 Perils

Having setup a bit of a foundation for you about the driving task, let's now get back to considering the nature of Level 3 and the perils is entails.

For an AI self-driving car of a Level 3, the SAE standard indicates that a human driver must be present during a driving journey and be ready for two major possibilities of driving action: (1) the human driver must be receptive to an AI system issued request for the human to intervene, and (2) the human driver needs to be aware of the status of the self-driving car such that if there is a need for the human driver to intervene due to a vehicle system failure that then the human driver will be ready and able to do so (regardless of whether the AI informs the driver to do so, since it might be that the AI does not inform the driver but that the human driver is supposed to take overt action on their own anyway).

I realize there's a lot packed into the definition for a Level 3. So, let's consider some examples.

One example provided by the SAE standard involves the aspect that suppose the radar sensors on a self-driving car fail or falter and suppose that the AI realizes this has happened. The AI would alert the human driver and ask that the human driver take over the driving task. It is open-ended as to how the AI would notify the human driver about this matter and each of the auto makers and tech firms are devising their own means of how to alert the human driver as to such matters.

It is not necessarily the case though that the AI might even realize that something is amiss on the self-driving car. Another example in the SAE standard consists of a tie rod that breaks during a driving journey. Would the AI realize that the tie rod has broken? Maybe, if the AI has the kind of sensors on the self-driving car that might detect such an anomaly. But, it could be that the self-driving car does not have sufficient sensors to detect this particular issue.

Suppose the human driver in the self-driving car could "feel" that the self-driving car was driving in a rather poor manner, possibly pulling hard to one side or the other of the road. The human might not know that the tie rod in particular is broken, but overall might realize that the car has suffered some kind of malfunction or breakdown. As such, the human driver might need to make a choice, do they continue with the AI driving the self-driving car, or would it be better to disengage the AI and have them take over the driving of the self-driving car?

This is a bit of a conundrum for the human driver.

Which is better, for the human to continue to allow the AI to drive the self-driving car, or for the human to instead takeover the driving of the self-driving car? The human doesn't know for sure what is wrong with the car. The human is unsure whether the AI can handle the driving of the car at this juncture. It is a bit disconcerting that the AI hasn't even apparently realized that the car is not as drivable as it once was.

Imagine you were in an Uber or Lyft, and the ridesharing car was being driven by a human driver. You are the passenger. All of sudden, the car begins to lurch somewhat and you can readily discern that something is amiss with the car. You look expectantly at the Uber or Lyft driver and wait to see what they are going to do about it. You'd be quite surprised that the Uber or Lyft driver seems to be completely unaware that the car is now lurching. Say what, the driver doesn't know that the car is in trouble? That is a scary proposition!

Well, that's exactly what can happen with a Level 3 self-driving car. The AI doing the driving might not be aware that there is something amiss with the car. Meanwhile, you, the human driver, sensing that there is something wrong with the car, need to make what could be a potentially life-or-death kind of choice.

If you decide to disengage the AI, you are hoping and betting that you'll be able to do a better job of driving the self-driving car than would the AI in this circumstance. If you decide to not disengage the AI, you are sitting there in fear that at some moment the AI is going to get completely thrown for a loop by the problem brewing and it might be that the AI will make a wrong decision and get things into an even worse predicament.

In short, we have the instance of the human driver having to be ready to respond and takeover the driving if the AI asks the human driver to intervene. But, this is not so easy as it might seem. How will the AI identify what is wrong and thus the reason for having the human driver take over? The human driver is expected to be able to take over in a timely fashion, but what is the nature of the time involved? What kind of action should the human driver take once they have taken over the driving task (this is entirely up to the human at that point).

We have the other instance whereby the human suspects that there is something amiss with the self-driving car and so the human has to decide whether or not to overtly and on their own opt to take over the driving task. The human driver might become concerned and even confused that the AI itself has not requested the human driver to take over the driving. This might imply to the human driver that they themselves are maybe falsely feeling that something is amiss, and the human driver might shrug off something serious, wasting precious moments that might have made a difference in taking over the driving of the self-driving car.

With the Level 2 self-driving cars, by-and-large most human drivers get the idea that they as a human driver need to remain well-connected and engaged in the driving task (I'm not saying everyone does this, just the preponderance seem to do this).

With the Level 3 self-driving cars, we are upping the ante in that the human drivers will be potentially lulled into assuming that the AI is able to handle the driving task (since it is more able to do so than with the Level 2), and thus the human drivers won't be ready when the AI either asks them to intervene or they won't be ready when on their own they should take over the driving task due to some kind of vehicle system failure.

You might say that in my Figure 2 we are going from the left side of the crossover point, consisting perhaps of Level 2 self-driving cars, and will now with Level 3 we will be entering into the right side of the crossover point. We are entering into the zone of attention deficits. The danger zone.

Devices for Driver Attention Detection

Some believe that the solution for potential attention deficits of human drivers consists of using various attention detection devices.

In a self-driving car, the steering wheel might be augmented by a sensing device that would ascertain whether the human driver has their hands on the wheel. If the driver seems to not have their hands on the steering wheel, the steering wheel might light-up to remind the driver to put their hands back onto it, or there might be a buzzing sound or some other means to alert the human driver about needing to keep their hands on the wheel.

The means of prompting the driver could include audio alerts, visual alerts, and tactile alerts (any variation thereof, or possibly all three modes at once).

Another sensing mechanism might be to have a camera mounted in front of the driver and facing toward the driver.

The camera might keep track of the driver's head, being able to detect if the head of the driver appears to be tilted or turned away from looking forward at the road ahead. This might also include a capability to detect eye movements. Thus, even if the head is aiming forward, it could be that the eyes of the driver are not also focused on the road ahead, and so the facial detection aspects of the eye movement could be another element for tracking. As per the steering wheel rejoinders, if the camera detection determines that the driver does not seem to be focused ahead, it will emit a sound or light-up an alert to jog the driver into compliance.

As a penalty for a driver that seems to repeatedly be lax in their physical attention forward, the self-driving car might either restrict the driver from being able to drive or the AI might opt to perform a fallback operation to bring the car to a minimal risk condition such as pulling over and parking at the side of the road. Though this seems like an appropriate precaution, it is not risk free as the driver might otherwise rebel against the system or take untoward action as a result of this kind of monitoring and system action.

Of course, we all know that distracted drivers are a problem on our roadways. The advent of smartphones has seemed to exacerbate the problem of drivers that are tempted to try and do two things at once. The driver wants to look at the latest tweet and at the same time be driving the car. If the roadway appears to be of a monotonous nature in terms of traffic, the driver figures that they can handle both the reading of their chats and the driving of the car.

Distracted driving is not limited to simply looking at your smartphone while driving.

A distracted driver can be focusing on a crying baby in the backseat of the car and be therefore no longer focused solely on the driving task. I've seen drivers that were engrossed in a heated debate with a passenger in the front seat of their car and had become so preoccupied with the acrimonious discussion that they no longer knew what was happening on the road ahead of them. There are drivers that put their make-up on while heading to work, and they too are distracted. There are drivers that are looky-loos that cannot help but look at a disabled car on the side of the road, whisking past at 80 miles per hour, and having their heads turned and looking away from the upcoming traffic. Etc.

Let's though clarify the difference between a so-called "distracted" driver and an inattentive driver.

In the case of a distracted driver, there is some other activity or aspect occurring that has drawn their attention. It could be that the driver themselves have opted to focus on the distractor, or it could be that the distraction itself has spurred the driver to become distracted. Either way, we can say that the driver has their attention elsewhere other than solely the driving task.

Thus, there can be a driver that has become inattentive, and the basis for the inattentiveness is due to distraction to something else.

We can also have an inattentive driver that is not necessarily "distracted" in the purist sense of distraction. An inattentive driver might be looking straight ahead and seemingly focused on the road, and not be talking with anyone and not be doing any other apparent activity, and yet they could still be inattentive to the driving task. Their mental state is the final determiner of whether or not they are providing attention to the driving task.

In case you think I am splitting hairs on this point, I am not. It is crucial to understand that an inattentive driver is not always and nor necessarily a distracted driver (if by "distracted" we mean there is some other physical manifestation that has drawn away the attention of the driver, which is the usual meaning most commonly applied).

Take a look at Figure 6.

Eliot Framework: Driver Attention Matrix

Figure 6

Physical Attention

Is

	Mentally Disengaged *(Awareness Drift)* 02	Being Attentive *(Prescribed Attention)* 01 ⭐
Is Not	Dangerously Inattentive *(Task Disengagement)* 04	No Eyes On The Road *(Controls Accessibility)* 03
	Is Not	Is

Mental Attention

Mental Attention
Physical Attention

01: Prescribed Attention = Mental + Physical
02: Awareness Drift = Physical – Mental
03: No Eyes on Road = Mental – Physical
04: Dangerously Inattentive = 0 *(No Mental, No Physical)*

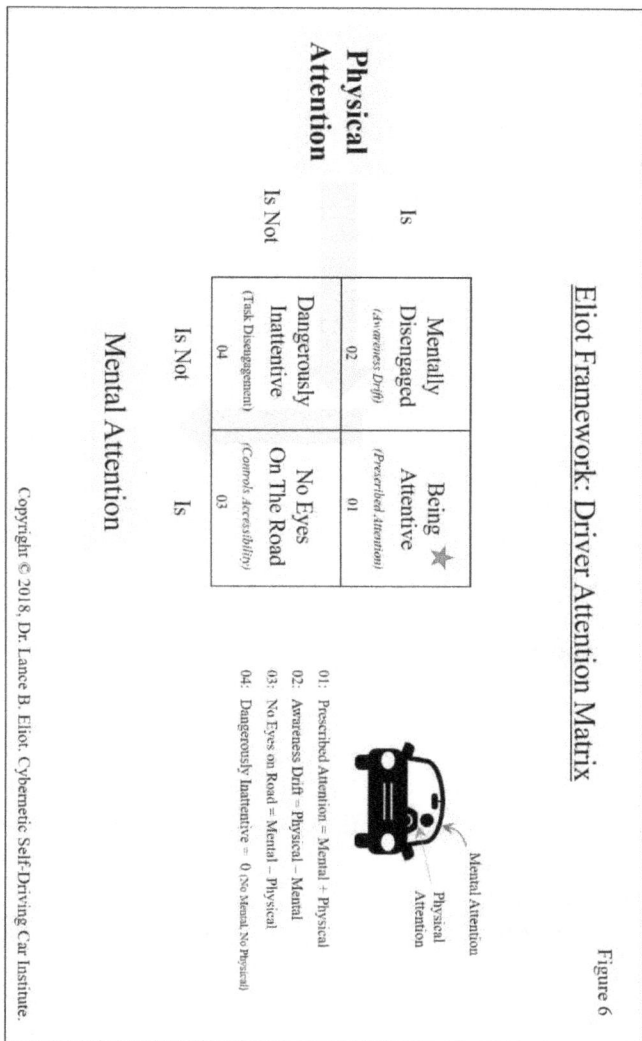

Copyright © 2018, Dr. Lance B. Eliot, Cybernetic Self-Driving Car Institute.

In the Driver Attention Matrix, there consist of rows that constitute the physical attention provided by the human driver and there are columns that represent the mental attentiveness of the human driver.

Suppose the human driver is physically aligned with watching the road ahead. Our steering wheel sensor says that the hands of the driver are present on the steering wheel, and the camera says that the driver is facing forward, and their eyes are on the road. Does this physical positioning and posture of the human driver then guarantee that they are attentive to the driving task?

No, it does not.

These handy gadgets to detect and warn the driver about being attentive are primarily dealing with the physical aspects of the human driver. By-and-large, they will catch the "distracted" driver that is looking down at their smartphone or turning their head to coo at the baby in the back seat. The hope is that by potentially preventing the driver from physically being inattentive, they will realign themselves physically and then presumably (hopefully) put their mind into the game too.

There is certainly a much greater chance of the driver being attentive to the driving task when their body is aligned with the needs of the driving task. Keep in mind though that it is somewhat like the proverbial line that you can bring a horse to water but you can't necessarily make it drink. Just because you force the driver to be physically "attentive" it does not ergo bring their mind to the matter as well.

The Driver Attention Matrix shows the four different possibilities of the circumstance when a driver is physically attentive versus not physically attentive, and combined with the driver's aspects of being mentally attentive versus not being mentally attentive.

The optimum or "best" driving setting is when the human driver is both physically attentive and mentally attentive (see the Matrix square labeled as "01").

When the driver is physically attentive but not mentally attentive (see the Matrix square labeled as "02"), you have a driver that is not mentally engaged in the driving task. This mental inattentiveness can mean that even though their hands, feet, and eyes are in the right spots to react to a driving urgency, their mind is not ready. The lack of mental attentiveness can therefore create delays in responding on a timely basis, plus their mental drift might prevent them from even knowing that action is needed and nor what should be the proper action to take, and might overall negate the aspect that their body was at least in a position to take needed action.

If the driver is mentally attentive but their physical attentiveness is not properly in place (see the Matrix square labeled as "03"), this lack of "eyes on the road" can mean that their mind won't even get a chance to determine that is something amiss.

If their hands and feet are also misplaced, the mind that tries to command their body to take action will have likely delays in physically getting into place in-time. Plus, the act of trying to get their body properly in place could cause other adverse consequences, such as steering in the wrong direction or jamming on the accelerator when they meant to hit the brakes (this happened to me when an elderly driver got confused and rear-ended my car at a stop when he inadvertently put his foot on the accelerator rather than the brake pedal).

The most dangerous driving attention deficit consists of the Matrix square labeled as "04" and involves a human driver that is both physically inattentive and mentally inattentive.

Today's devices that try to detect when a driver is "distracted" from the driving task are dealing with the physical attentiveness aspects. This is helpful, but it is only part of the story. It is one-half of the coin. As mentioned before, the horse still needs to drink the water once they have been forced to the trough.

The emergence and advancement of the various physical attentive detection devices are really just a surrogate for deriving the mental attentiveness. It is yet unknown how much of the time this works in that regard of also forcing someone to mentally reengage in the driving situation.

I'd suggest that we cannot say that it is 100% of the time that a physically inattentive driver that was sparked into becoming attentive is also going to become a mentally attentive one too. As such, if the physical attentiveness gets us to say mental attentiveness 90% of the time, it still bodes for concern that 10% of the time the driver is mentally still inattentive. Or, it could be 80% and 20%, or maybe even 50% and 50%. Or worse.

You might be wondering how we could ascertain whether a human driver is mentally attentive to the driving task?

It's a hard nut to crack. There is no particular means to somehow use sensors to detect that the human is actively thinking about the driving task, though in the future there are some that predict that we might able to do so via brainjacking.

Another approach could consist of having the AI converse with the human driver and try to determine whether the driver is aware of the driving situation at-hand.

This would be somewhat akin to when you are helping a novice teenage driver to drive a car. The novices sits in the driver seat, and you try to coach them as to what is taking place. This is a kind of co-sharing arrangement of the driving task, though you usually don't have your own driving controls and must rely upon the teenager to undertake the maneuvering of the car.

If you've ever done this kind of driving assist, you know that part of the time you are gauging whether the teenager is aware of the driving situation, and at other times you might be telling them what to do or urging them to take certain actions. The more savvy they become, the less you are likely to offer commands and instead just probe their mental state to ensure that they know there are pedestrians nearby or that they are getting rather close to the car ahead of them.

The AI would potentially be able to ascertain whether the human driver is mentally ready or engaged and is indeed ready to undertake the driving task if needed. The AI could periodically interact with the human driver. If the physical attention sensors detect that the human driver is physically becoming inattentive, the AI could then be part of the alert for the driver and also then act secondarily to try and ensure that the driver mentally gets back into the appropriate driving mode too.

This is no silver bullet, unfortunately. First, the AI would need to be good enough to be able to carry on such a dialogue and be "smart" enough to judge whether the human was immersed in the driving moment or not. Second, the act of the AI conversing with the human driver can itself be considered a form of potential distraction. Third, the method of communicating, if verbal, might be too slow and crude to effectively convey in a timely fashion whatever urgency might be arising. And so on.

Reframing the Standard Levels

By now, I hope that you are grasping the concerns associated with the Level 3 self-driving cars. It is somewhat ironic that though they are more capable than a Level 2, it also tends to cause the human driver to become more inattentive, which then gets us into the perils of Level 3.

If we get a lot of Level 3 self-driving cars on our roadways, what might happen?

Well, we could see a lot of unfortunate car crashes that injure or kill people. This could happen during that boundary time of when the AI is trying to get the human driver to intervene, but the human driver is seemingly caught unaware. Or, the AI or the self-driving car has experienced some kind of failure, but the human driver did not take over the driving controls of their own volition and therefore the situation took a turn for the worse.

The auto makers and tech firms that make the Level 3 self-driving cars would undoubtably try to cast these incidents as a human failing. It wasn't the fault of the AI per se, they would contend, it was those human drivers that were lazy, distracted, inattentive, confused, or whatever. Even if this could be used as a defense, the end result of having injuries and deaths is going to be enough to likely put the kibosh on the self-driving desire that seems right now to be rolling forward. One could expect a public backlash along with a regulatory backlash.

Unfortunately, the odds are that if the Level 3 self-driving cars become the bad apple in the barrel, it might well spoil the rest of the barrel. If the public and regulators perceive that Level 3 is an "autonomous" self-driving car, they will naturally extend their concerns to the Level 4 and Level 5 self-driving cars. All self-driving cars will be tossed into the same bucket, regardless of their capabilities.

Take a look at Figure 7.

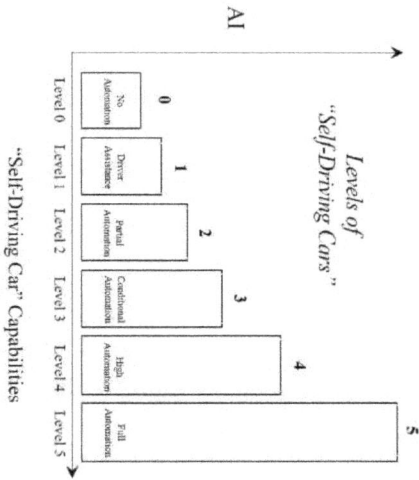

Eliot Framework: Levels of AI Self-Driving Cars

Figure 7

Levels of
"Self-Driving Cars"

| 0 | 1 | 2 | 3 | 4 | 5 |

| No Automation | Driver Assistance | Partial Automation | Conditional Automation | High Automation | Full Automation |

Level 0   Level 1   Level 2   Level 3   Level 4   Level 5

"Self-Driving Car" Capabilities

Copyright © 2018, Dr. Lance B. Eliot, Cybernetic Self-Driving Car Institute.

This diagram indicates the six levels of self-driving cars as ranging from a Level 0 to a Level 5. I've tried to indicate that the AI capability increases as the levels get higher in number.

Some would argue that the Level 4 and Level 5 should be much higher and tower well above the levels below Level 4, doing so to signify how much an improvement in self-driving there is after you get past Level 3. I wanted to keep the diagram readable and so I have shown the Level 4 and Level 5 as at least somewhat taller than Level 3. If you like, think of them as a lot taller.

One of the concerns about the levels is that they are all part of the overall definition associated with self-driving cars. But, the reality is that the Level 0, 1, 2, and 3 are not really truly autonomous cars, and yet they are being associated overall with self-driving cars.

Take a look at Figure 8 (see next page).

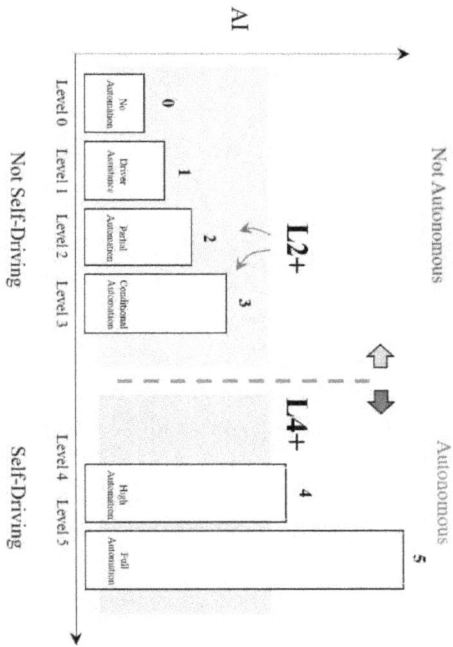

Eliot Framework: Bifurcation of Autonomous Driving Levels

Figure 8

Copyright © 2018, Dr. Lance B. Eliot, Cybernetic Self-Driving Car Institute.

As shown, I've divided up the levels into two major groupings. The Level 3 and the levels below it are all considered as not autonomous. The Level 4 and Level 5 are considered to be autonomous.

This is a bifurcation of the autonomous driving levels.

By bifurcating or splitting the levels into two groups, we can then perhaps differentiate between the two groups. It is a potential means of reframing the underlying nature of these cars.

This might help when dealing with any dilemmas associated with Level 3. If the public and regulators, and the media, began to realize that the Level 3 is in the "not autonomous" camp, it might be instructive when the time inevitably comes for them to have qualms about "self-driving cars" — since one might then clarify that the Level 3 is not actually a true self-driving car at all.

The Velodyne white paper that I earlier had mentioned in this matter has proposed that any vehicle that is less than Level 4 should not be construed as and nor referred to as an autonomous vehicle and nor a self-driving or driverless vehicle.

Furthermore, going even further, any such vehicle that is less than Level 4 should overtly and directly clarify that it is not an autonomous vehicle, nor a self-driving or driverless vehicle. By being overt about this matter, it would hopefully reduce the likely implied aspect that if it does not say that it isn't self-driving then perhaps one could infer that it is self-driving (this would likely be suggested by the marketing aspects, I'd wager).

And, instead of separately referring to Level 2 and Level 3, it has been proposed that those two levels be collectively referred to as Level 2+. Meanwhile, the Level 4 and Level 5 might be referred to as Level 4+.

Take a look at Figure 9.

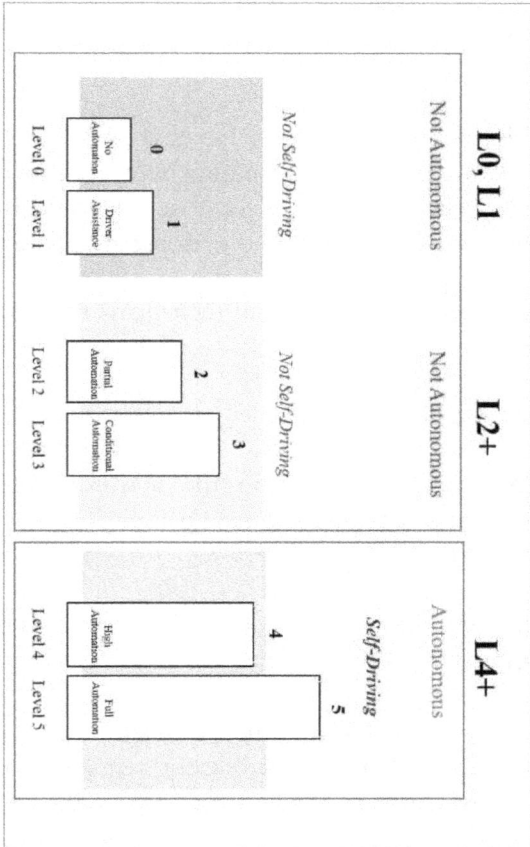

Eliot Framework: Bifurcation of Autonomous Driving Levels

Figure 9

As shown in the diagram, we would have a bifurcation of the levels into two major groups, consisting of the not autonomous levels and the autonomous levels.

Within the autonomous levels, they would be collectively referred to as L4+. For the not autonomous levels, the Level 2 and Level 3 would be referred to as L2+.

Presumably, the lowest levels, Level 0 and Level 1, which are included in the not autonomous grouping, would simply be referred to as Level 0 and Level 1 (though, I suppose we could try to come up with something catchy like say "L1-" or something like that).

Can we get the auto makers and tech firms to adopt this kind of nomenclature?

It will certainly be hard to do.

The jockeying for position in the autonomous car realm is tremendous and those that are marketing their wares will often go to rather sketchy lengths to do so.

Efforts to seek near-term glory need to be balanced with the future of the industry.

Auto makers and tech firms need to take a longer-term perspective and realize that if the public and regulators and media turn toward blocking progress in the self-driving car industry, it will hurt everyone involved.

By being more mindful about how we describe these innovations, it might provide some semblance of a chance to keep the entire barrel from being cast as no good.

This is not just trying to call a rose by some other name. This is differentiating what one kind of "automation equipped" car can do versus what a truly autonomous self-driving car can do. We need to be able to call an orange an orange, and an apple an apple. Right now, we appear to be heading toward a boiling point (recall the story of the frog!), and it would make sense to do something before it is too late and the pot boils over.

.

CHAPTER 9

CRYPTOJACKING

AND

AI SELF-DRIVING CARS

CHAPTER 9

CRYPTOJACKING AND
AI SELF-DRIVING CARS

Are you into bitcoins? Actually, I should ask whether you are into cryptocurrency, since the word "bitcoin" can refer to a particular kind of cryptocurrency or it can also be used as a generic reference to digital currencies, similar to how the word "cryptocurrency" is a generalized way of mentioning digital currency or so-called electronic cash. It's also like saying "Kleenex" and meaning facial tissue versus referring to the specific brand, or like saying "Xerox" and meaning to copy something or making a reference to the well-known copier and workstations firm.

As a quick primer about bitcoins overall and cryptocurrencies, unless you've been living in a cave for the last ten years or so (by the way, October 31, 2018 was the tenth anniversary of the emergence of cryptocurrencies, which some cleverly pointed out that Halloween seems an apt time due to the aspect that it is perhaps "frightening" as to what these electronic coins might do to our world economy!), you certainly must be aware that there has been a move afoot toward trying to have electronic currency that is not particularly connected to today's real-world currency. These newly made-up currencies are hoping that people will believe they have intrinsic value and be willing therefore to make what are simply numbers in the computer network into something worthy of using, saving, and spending.

It is as though I opted to tell the world that I have a new kind of currency, the Lancelot coin, and asked everyone to please consider this as real money. Some might be excited about this new currency and be willing to exchange real-world monies for it, such as agreeing that perhaps one US dollar is equal to one-half of a Lancelot coin, and maybe worth 1,000 in Yen.

People might be willing to barter goods and services using the Lancelot coin. The electrician that comes over to fix your home wiring might say that you can pay in Lancelot's or in US dollars, whichever you prefer. When your offspring mows a neighbor's lawn, the neighbor might pay your child in Lancelot's rather than quarters and dimes.

Since the Lancelot is only an electronic currency, and not paper-based or physically composed of coins, it is conducive to transactions in today's online world. You can just trade the Lancelot's via your smartphone or tablet or laptop. No need to deal with the laborious aspects of physical coinage and paper-based money. The electronic currency dispenses with the expensive overhead involved in the physical aspects of currency, eliminating the usual woes of storing physical money, transporting it, keeping it in your safe at home or wallet or purse, etc.

Instead, it is all just numbers stored electronically. Nifty!

The underlying framework that dominates the cryptocurrency emergence is known as blockchain. A blockchain is a data structure that can be used to house the made-up cryptocurrency. Software is then used to maintain the blockchain, keeping track of the cryptocurrency as people opt to make use of it. Via the blockchain data structure and software, you can essentially have an on-line ledger that tracks and records transactions. In that sense, you might almost think of the blockchain for a particular cryptocurrency as being akin to a clearinghouse for the cryptocurrency.

When your neighbor electronically transmitted Lancelot coins to your child for mowing their lawn, the blockchain ledger for the Lancelot cryptocurrency would have been updated on-line to indicate that your neighbor had signed-over some of their Lancelot coins to your child. Your child would now likely have an electronic wallet setup on say their smartphone and be able to see how much Lancelot coins they have, along with then using those Lancelot coins when desired. For example, perhaps an ice cream truck drove down the street after the mowing was completed and your child decided to pay for getting some refreshing ice cream with they newly earned Lancelot coins.

Usually, a cryptocurrency is all on its own. By this I mean that it is not somehow backed by a country that says the cryptocurrency is their official form of money. Instead, some private party or maybe a public entity will decide to bring forth a cryptocurrency and see whether the world will be interested in it. If the ice cream truck wouldn't take Lancelot's, and didn't consider it to be of value, this would undermine the act of your child trying to use the Lancelot's to buy ice cream.

Indeed, only if enough people are willing to accept a particular cryptocurrency as having value, it really is nothing more than numbers on a network. Only once people agree that they will consider it of value does the value then emerge. If nobody likes the Lancelot currency, it is likely doomed to ultimately fall by the wayside. On the other hand, if people like the Lancelot currency and the electrician is willing to use it and the ice cream truck is willing to use it, there might be enough of an impetus that it will become a valued form of currency.

This has brought forth a kind of wild gold-seeking pandemonium. If you can startup a blockchain with the Lancelot cryptocurrency and get others to adopt the Lancelot, you could potentially become a zillionaire overnight. All it takes is to get other people to believe in the electronic currency that you are offering to the world. As the original issuer of Lancelot's, I might set aside a whole bunch of Lancelot's for me, and then once everyone else believes that Lancelot's have value, I've got a handy stash of them.

Since a cryptocurrency is only as valuable as believed by people, it can at one moment be high in value and in the next moment plummet in value. This is why of the now hundreds and likely thousands of newly "minted" cryptocurrencies, some are pretty much worthless right now and others are worth a ton. For any given cryptocurrency, it could be positioned on a bubble that will burst at any time.

In some respects, the belief is that some cryptocurrencies are maybe "too big to fail," which is sometimes used as a comparison to social media. Why does Facebook have around 2.2 billion active users? Partially because it has momentum. The more people that join Facebook, the more other people opt to join Facebook in order to communicate with them. It just keeps getting bigger and bigger. Likewise, if you have a cryptocurrency that becomes popular, presumably enough other new people are attracted to it, and eventually it grows large enough to possibly be self-sustaining.

Could everyone tomorrow decide to stop using Facebook? Yes, they sure could. As such, everyday is another lucky day for Facebook as long as people continue to believe in using Facebook. The same goes with some of the cryptocurrencies. The popular ones are hoping that a critical mass will make the electronic coins resilient and stable.

Advocates of cryptocurrencies emphasize that the use of blockchain aids in ensuring that the electronic currency is kept secure. This is a vital aspect of anyone that wants to believe in using electronic currency. In contrast to a country that controls how much currency the country decides to put into circulation, and along with the many ways in which fraudulent currency is detected and controlled, the use of cryptocurrency has the same kinds of concerns and yet it is not being maintained by a country per se.

Maybe some clever hacker is able to create fraudulent Lancelot's and make use of them. This means that the Lancelot's that your child so laboriously earned are cheapened because someone else was able to make ones on their own.

This then takes us to the aspects of how cryptocurrencies are generated. The primary method for the issuance of any particular cryptocurrency is via what is called electronic mining. People that believe in the cryptocurrency are able to use computers to mathematically find a nonce. Essentially, a nonce is a arduous number to find and involves a time-consuming effort and computer processing consuming effort to determine. It is used with a hashed version of a block's content and then can be verified relatively readily by other such electronic miners, and thus allows then for a block to be added into the blockchain. So, it's a hard number to find, but once it is found it then is relatively easy for others to confirm it.

This is considered a proof-of-work kind of system. The electronic miners that succeed in finding the nonces get rewarded by earning some of the cryptocurrency. In addition, this overall method and mathematical approach is what helps the blockchain to be considered "immutable" and "secure" – though you need to keep in mind this is often touted as an absolute, but that's not really the case. If you had enough computer processing power, you can contend with such aspects (thus, the concerns that some have about the emergence of quantum computing and whether quantum computers might be so fast that what we consider today to be secure will not be secure in the future).

You might be thinking that you could make a lot of dough, or let's say electronic currency, by setting up your own electronic mining capability. It's actually very easy to do. You don't need to necessarily even write any programming code. There are off-the-shelf apps that will do the mining on your behalf. Thus, you really don't need any particular computer expertise per se and can pretty much get things setup as a miner and then just start mining. There's gold in them thar hills!

But, not so fast. You would need to first buy some computers to use for the mining effort. That's a cost for you. Or, you could rent the use of someone else's computers, but again that's a cost since you would need to pay for the rental usage. Thus, you need to consider the cost of the computer related equipment that will be doing the mining. There would be the initial setup costs and then any ongoing costs to keep those computers humming along and doing the mining.

What is actually the more significant cost, and somewhat surprising to many, involves the cost of the electrical power to run the computers that you are using for doing the mining. I know this seems odd. Most of us are used to plugging our computer into a wall socket and we don't think much about the cost of the electrical power to run the computer.

In the case of the mining, you would likely want to use high-end computers that are very fast, and that are likely electrical power hogs. You'd want to use not just one such high-end computer but perhaps dozens or maybe even hundreds, if you could. The more mining, the more chances of earning that miner's income.

Some prospective entrepreneurially minded miners originally opted to setup some computers in their home or apartment and figured that while they were at the office during the day and doing their daytime job, all of those computers at home were earning them extra "cash" by mining for a particular cryptocurrency.

When they saw the electrical power bill at the end of a month, they were shocked (there's a pun for your day!). Between the cost of getting and maintaining the computers, along with the whopping electrical power bill, it turns out that mining wasn't necessarily a slam dunk ROI (Return on Investment). Instead, the amount of cryptocurrency that could be earned, along with its fluctuating value, and offset by the costs of the mining effort, turns out that it was not especially profitable for some miners.

Indeed, there has been a trend of people that want to mine for cryptocurrency to find a place to live and work that happens to have low or lower electrical power costs.

The U.S. national average residential cost for a kWh (kilowatt hour) of electricity Is about 7 cents. Some estimates are that to mine the typical cryptocurrency you might need to consume 215 kWh of electricity. The state of Washington has an average residential cost of 4.37 cents per kWh, which is the lowest in the nation (followed by Montana, Oklahoma, Louisiana, etc.). Thus, some miners have setup their "mining camp" in those lowest electrical cost states.

Another way to mine might be to tap into someone else's computing and use their computers for your own purposes of mining. Assuming that the miner does so without the permission of whomever owns those computers, this is actually a type of crime known as cryptojacking.

Cryptojacking consists of making use of someone else's computers to try and mine for cryptocurrency. The bad-hat hacker or cyber criminal would do this to avoid their own costs of doing the mining and leverage your "free" computing power for their own gain. Rather than having to deal with setting up computers and paying for the electrical power, instead the hacker somehow gets you to unwittingly provide your computers for their fiendish purpose.

When I tell people about the latest trend of cyberhacking involving cryptojacking, it is interesting that they often ask me whether the miner is "harming" the computer in doing so. In other words, they wonder whether the use of the mining will adversely impact their computer, assuming that a miner was somehow able to turn the computer into a miner for them.

Let's start with emphasizing that the miner is stealing your computer processing for their own ends. In that sense, they are taking a kind of good or service away from you. If you own that computer, you presumably own the processing cycles available on it. If someone else is using it for free and without your permission, they are stealing something of value, whether or not you happened to be using it. I liken this to someone that takes your car at night while you are asleep. Sure, you might not be using it while you are asleep, but I think we would all agree that it is nonetheless untoward that someone is taking your car at night without your permission and using it.

Assuming that a miner has been able to coopt your computer for mining, they might do so sneakily and you would not even realize that your computer is being used for this purpose. The mining could take place whenever there are idle or available processing cycles, and meanwhile the computer seems to be working just as normal. Of course, the thief will need to get access to the computer to see what the mining results are, and so in that sense they would have to have other unfettered online access to your computer too.

The mining effort might be setup to use up a lot of your computer processing. Thus, when you try to use your computer to write that great novel or need to play some games, your computer might seem to work in a halting fashion. It could be that the mining effort is chewing up computer processing and setup as a higher priority on your computer than the other tasks that you want it to do. A sluggish or slow responding computer could be the result of the mining.

This also assumes that the miner is only interested in using your computer for mining. I don't want you to fall into the mental trap of some that say it is a "harmless" kind of act. Keep in mind that whomever has cracked into your computer is now likely able to do other untoward acts. They might decide to ultimately do a ransomware effort on your computer, perhaps once they believe you are realizing that mining is occurring, and you are going to try and cutoff the mining.

The odds are that anyone that would be willing to break into your computer to do the mining will likely have other nefarious thoughts and acts to be played out. They might grab up your private info and try to use it for identify theft. They might do all sorts of horrendous acts, especially when they think the gig is up and it is time to move on to someone else's computer for mining.

Anyone that somehow tries to suggest that cryptojacking is a "victimless" crime is not thinking straight. Besides the criminal act of stealing your computer processing, it is a quick and easy slippery slope of then going after other nefarious aspects of doing things to your computer and likely stealing and exploiting whatever you have on that computer.

When I refer to your computer, I not only mean a desktop computer that you might have at home, but also your laptop that you take on the road with you, and even your smartphone or tablet can all be cryptojacked.

You can consider the crytojacking to be akin to a computer virus. If your computing device gets such a computer virus on it, your computer then becomes a zombie soldier in the army of the miner.

As with any computer virus, you should have up-to-date cyber protection software running on your computing devices to hopefully prevent the cryptojacking and at least detect it if indeed your computer has gotten infected. Once infected, you'll want to carefully take steps to have the mining virus removed and also figure out whether the miner has done anything else untoward on your computer. It can be a terrible moment of reckoning once you realize that a miner has infected your computer.

What does this have to do with AI self-driving cars?

At the Cybernetic AI Self-Driving Car Institute, we are developing AI software for self-driving cars. One aspect for AI self-driving cars is the potential for them to be exploited for cryptojacking purposes.

Let's consider why an AI self-driving car might be an attractive target for cryptojacking.

AI self-driving cars are going to be chockfull of high-end computer processing capabilities. It takes a lot of computing to be able to handle all of the AI aspects involved in driving a car. In addition, the computers on-board the self-driving car are dealing with many other elements of the car, including its numerous subsystems for engine control, drivetrain control, etc.

The odds are too that AI self-driving cars will be filled with elaborate infotainment systems. It is anticipated that AI self-driving cars will likely be running around the clock, providing ridesharing services, and the odds are that passengers will be seeking on-board in-car entertainment during potentially lengthy rides (your daily commute to work of two hours will no longer require your attention to the driving task, and so the odds are that you might want to watch movies, maybe do online educational courses, etc.).

Thus, for a cryptocurrency miner, the computing capabilities of an AI self-driving car are quite attractive. If you could somehow load a mining virus onto the on-board system of the self-driving car, it could crank away at doing mining. This mining could happen while the AI self-driving car is perhaps parked in your garage at home and recharging. Or, it could happen while the AI self-driving car is cruising around the neighborhood or when it is on its way to a destination.

Similar to the aspects of doing mining on your smartphone or laptop, the mining activity could be purposely hidden from view. The mining virus might only tap into the computers when it figures that doing so will not otherwise undermine the operation of the self-driving car. In that sense, it can remain hidden and you might not even realize it is there.

Or, the mining might become disruptive to the on-board systems. This can have potential dire consequences. If the mining consumes processing that was supposed to be determining whether the AI self-driving car should hit its brakes or make that left turn up ahead, it could be a life-or-death kind of distraction to the AI and have terrible consequences.

There's also the same concern that if the mining virus is actually able to do mining, it tends to suggest that the virus implant and the bad-hat hacker might be able to do other untoward acts to your AI self-driving car too. They might be able to decide where to have the AI self-driving go. Maybe they opt to use the self-driving car whenever they wish to do so. Maybe they opt to have the AI self-driving car carry out bad acts. Etc.

For most of the auto makers and tech firms that are developing AI self-driving cars, they consider this notion of potential cryptojacking to be an edge problem. An edge problem is a corner case or considered at the edge of the core aspects that you are trying to solve. Right now, the auto makers and tech firms are focused on getting an AI self-driving car to do the normal things that you would want a self-driving car to do, such as driving down the road and not hitting anything along the way.

I won't get into a debate herein about whether or not cryptojacking is an edge problem per se.

On the one hand, there are so few AI self-driving cars as yet that it would seem somewhat unlikely that computer criminals are already plotting to use AI self-driving cars for this purpose. Furthermore, the odds are that computer criminals are aiming to use more conventional computer viruses against AI self-driving cars, and the notion of doing cryptojacking is a lot lower on their list of nefarious acts.

Also, most pundits are assuming that the AI self-driving car industry is already preparing for any kind of computer viral attack. As such, there is presumably no need to be concerned about any one particular type of such attack. Just brace the AI system to be able to detect and cope with any kind of computer virus, and therefore specific ones like the cryptojacking should get detected and stopped. At least, that's the theory of it.

I'd like to clarify and introduce the notion that there are varying levels of AI self-driving cars. The topmost level is considered Level 5. A Level 5 self-driving car is one that is being driven by the AI and there is no human driver involved. For the design of Level 5 self-driving cars, the auto makers are even removing the gas pedal, brake pedal, and steering wheel, since those are contraptions used by human drivers. The Level 5 self-driving car is not being driven by a human and nor is there an expectation that a human driver will be present in the self-driving car. It's all on the shoulders of the AI to drive the car.

For self-driving cars less than a Level 5, there must be a human driver present in the car. The human driver is currently considered the responsible party for the acts of the car. The AI and the human driver are co-sharing the driving task. In spite of this co-sharing, the human is supposed to remain fully immersed into the driving task and be ready at all times to perform the driving task. I've repeatedly warned about the dangers of this co-sharing arrangement and predicted it will produce many untoward results.

Let's focus herein on the true Level 5 self-driving car. Much of the comments apply to the less than Level 5 self-driving cars too, but the fully autonomous AI self-driving car will receive the most attention in this discussion.

Here's the usual steps involved in the AI driving task:

- Sensor data collection and interpretation

- Sensor fusion

- Virtual world model updating

- AI action planning

- Car controls command issuance

Another key aspect of AI self-driving cars is that they will be driving on our roadways in the midst of human driven cars too. There are some pundits of AI self-driving cars that continually refer to a utopian world in which there are only AI self-driving cars on the public roads. Currently there are about 250+ million conventional cars in the United States alone, and those cars are not going to magically disappear or become true Level 5 AI self-driving cars overnight.

Indeed, the use of human driven cars will last for many years, likely many decades, and the advent of AI self-driving cars will occur while there are still human driven cars on the roads. This is a crucial point since this means that the AI of self-driving cars needs to be able to contend with not just other AI self-driving cars, but also contend with human driven cars.

It is easy to envision a simplistic and rather unrealistic world in which all AI self-driving cars are politely interacting with each other and being civil about roadway interactions. That's not what is going to be happening for the foreseeable future. AI self-driving cars and human driven cars will need to be able to cope with each other.

Returning to the topic of cryptojacking, let's consider some additional facets about this potential malady.

First, you might be wondering how would the mining virus somehow work its way into the on-board systems of the AI self-driving car. Turns out that there are various opportunities for an infection to gain entry and spread within an AI self-driving car.

One approach would be for someone to get physically into the AI self-driving car and plug-in a dongle into the OBD-II port. This would be relatively easy to do in that if your AI self-driving car is available for ridesharing purposes, a nefarious person could simply book a ride and then plug in the dongle. This is why I've previously called for either disabling the OBD-II port or adding hefty security capabilities to it and its use.

If you've ever seen the ads by Progressive Insurance and other car insurers, you've perhaps seen that they offer a discounted insurance rate if you are willing to allow them to track your driving usage. They do so by having you plug-in a dongle into the OBD-II port that usually sits just under the dashboard of your car. Via the port, the dongle is able to collect driving related data and can electronically transmit it to the insurer (or, collect the data onto the dongle and then have you physically provide the dongle back to the insurer).

The OBD-II port not only allows for collecting data from the car, but you can also use it as an input into other allied subsystems of the car. Thus, a miner with a dongle that has a pre-loaded infectious mining program could potentially slip the virus into your AI self-driving car. Presumably, hopefully, the various on-board systems security systems would detect and prevent this from happening, but its an ongoing cat-and-mouse game as the hackers find new ways to break through security screening.

Another method of infection could be via the use of the Internet of Things (IoT). As I've mentioned previously, the odds are that AI self-driving cars will have a slew of IoT devices. Some of the IoT devices will be included by the auto maker or tech firm that makes the AI self-driving car. Some of the IoT devices will be provided by third parties as add-on capabilities for your AI self-driving car. And, of course people getting into the AI self-driving car will likely have IoT devices on them or with them, which will also be communicating with your AI self-driving car.

A bad-hat hacker might load-up an IoT device that otherwise seems legitimate and yet has a cryptocurrency mining element on it. This IoT device then comes in contact with your AI self-driving car and perhaps has some clever means of infusing itself into the on-board systems.

Another approach of infection could be via the OTA (Over The Air) capability. AI self-driving cars are equipped with an electronic communication for uploading data from the self-driving car into the cloud of the auto maker or tech firm. This OTA also allows for pushing down into the AI self-driving car the latest patches and updates of the AI software. If a bad-hat actor could somehow plant the mining tool into the OTA patches or updates, it could ride like a Trojan Horse into the guts of the AI self-driving car system.

As was widely reported in the media earlier this year, a bounty hunter "white hat" security firm discovered that Tesla had some of their Amazon Web Services (AWS) cloud instances infected with a cryptocurrency miner. Tesla indicated that it was confined to only their internally-used engineering test cars and otherwise had no material impact.

One aspect worth noting is that miners especially look to hide their mining activity by trying to target large companies rather than someone's individual at-home PC, since the odds are that the added use of computing and the added electrical power consumption will not be noticed by a large firm and also they can tap into a lot of processing power all at once.

In any case, beyond the means noted in this discussion, there are even additional potential points of entry that a computer virus can potentially make its way into an AI self-driving car. The auto makers and tech firms need to be mindful of the importance of systems security and provide robust protections accordingly.

In theory, the AI self-driving car should have sufficient protections to prevent a cryptojacking virus from getting into the on-board systems. But, suppose somehow that one managed to wean its way in. What then?

The AI of the self-driving car needs to have sufficient self-awareness to realize that the cryptojacking is embedded and running. Doing so involves performing various internal systems checks to be assured that what is running on-board is what is intended to be running. Also, the Operating System (OS) used for the AI self-driving car needs to be savvy enough to be judging the load balancing of the processors and making sure that some unknown "hog" is not needlessly chewing up processing cycles.

Imagine a cryptojacking that might somehow be able to get implanted into a multitude of AI self-driving cars, perhaps an entire fleet, all of which might be then used for mining by the bad-hat actor. Assuming that the bad-hat thief does nothing else nefarious (unlikely!), just envision the huge amount of processing capability that could be used for mining. Breathtaking!

On a related topic, let's turn around on this illegal cryptojacking notion and consider the other side of the proverbial coin, namely a legitimate version of mining. Suppose you owned an AI self-driving car and you wanted to intentionally use it for doing cryptocurrency mining. The processors on-board are your processors, so why shouldn't you be able to leverage their availability?

Sure, you would not want the mining to inadvertently undermine the processing needed by the AI to drive the self-driving car. As such, you would want this to be done in a very methodical and calculated way. The AI self-driving car would need to have a capability to make use of the "available" processing, such as when the self-driving car is at idle or maybe parked in your garage at home. Otherwise, it would allow the processors to be used for things that are benign, assuming that the mining software you use is well-contained and well-behaved.

When I mention this possibility at AI self-driving car industry events during my presentations, you can imagine the collective groan that comes from the AI developers there. What, we have to now add to the AI self-driving car a feature to allow the processors to be used for something other than driving the car? Are you crazy? We have our hands full just being able to get the AI to drive the car. Also, they rightfully have concerns that the mining might go astray and detract from the AI driving task.

I realize that the idea of using the on-board processing for anything other than the necessary aspects of driving the AI self-driving car is rather hard to envisage right now. Until we have perfected AI self-driving cars and have had them working appropriately, trying to put any time or attention toward a rather ancillary aspect as leveraging the processing cycles seems wholly unimaginable.

I'd bet though that eventually there will be a sense of stability that will then open the possibility of harnessing the incredible "extra" processing power of the AI self-driving cars. At that future time, some enterprising owners of AI self-driving cars will likely begin to think about how else they can possibly make money from their AI self-driving car, beyond using it as a ridesharing service.

There are also going to be some people that might be thinking beyond the money-making side of things. For those of you that have ever aided the search for extraterrestrial intelligence via the SETI program, you likely know about the free program that you can download onto your home computer and use it to help analyze radio telescope signals. When I tell people about this aspect, they at first think it seems kind of nutty. Does this mean that you have to believe in Martians and little green men? No, it is just a fun and interesting way to contribute toward the search for what might be on other planets.

In that sense, you could use the available processing time of your AI self-driving car to help find intelligent life somewhere else in our universe. Though that might seem somewhat farfetched as a reason to use the extra processing capability, the point is that beyond just using the processing for possibly mining of electronic currency, you could potentially use that processing for lots of other purposes, some worthwhile and some perhaps a bit out-there.

There are numerous mathematical puzzles that are starving for computer processing time to be solved. As well, there are lots of Machine Learning (ML) aspects that require gobs of processing time. You could potentially use an AI self-driving car for those purposes. You might even join with friends that also have AI self-driving cars and you all agree to each make use of your respective AI self-driving cars for some in-common endeavor.

It is also notable to consider that with an entire fleet of AI self-driving cars, the collective "extra" computational processing could be substantive, and especially too if you were to consider this processing available at differing magnitudes throughout a 24x7 non-stop period of time. As mentioned before, the potential for this use must also be balanced by the amount of energy consumed, along with ensuring that the AI self-driving cars are able to carry out their primary function unabated.

Will your AI self-driving car become a slave to a cryptocurrency miner?

Let's hope not.

Could you possibly use your AI self-driving car's processing capabilities to provide some kind of added service beyond being just a self-driving car?

Maybe so, though likely far off in the future.

For now, let's focus on keeping those nefarious crypto miners from undermining AI self-driving cars and make sure that we put in place sufficient cyber security guards that the bad-hats won't be able to setup camp in your self-driving car. That's a triumphant accomplishment that is worthy of a loudly proclaimed Eureka!

APPENDIX

APPENDIX A
TEACHING WITH THIS MATERIAL

The material in this book can be readily used either as a supplemental to other content for a class, or it can also be used as a core set of textbook material for a specialized class. Classes where this material is most likely used include any classes at the college or university level that want to augment the class by offering thought provoking and educational essays about AI and self-driving cars.

In particular, here are some aspects for class use:

o   Computer Science. Studying AI, autonomous vehicles, etc.

o   Business. Exploring technology and it adoption for business.

o   Sociology. Sociological views on the adoption and advancement of technology.

Specialized classes at the undergraduate and graduate level can also make use of this material.

For each chapter, consider whether you think the chapter provides material relevant to your course topic. There is plenty of opportunity to get the students thinking about the topic and force them to decide whether they agree or disagree with the points offered and positions taken. I would also encourage you to have the students do additional research beyond the chapter material presented (I provide next some suggested assignments they can do).

RESEARCH ASSIGNMENTS ON THESE TOPICS

Your students can find background material on these topics, doing so in various business and technical publications. I list below the top ranked AI related journals. For business publications, I would suggest the usual culprits such as the Harvard Business Review, Forbes, Fortune, WSJ, and the like.

Here are some suggestions of homework or projects that you could assign to students:

a) <u>Assignment for foundational AI research topic</u>: Research and prepare a paper and a presentation on a specific aspect of Deep AI, Machine Learning, ANN, etc. The paper should cite at least 3 reputable sources. Compare and contrast to what has been stated in this book.

b) <u>Assignment for the Self-Driving Car topic</u>: Research and prepare a paper and Self-Driving Cars. Cite at least 3 reputable sources and analyze the characterizations. Compare and contrast to what has been stated in this book.

c) <u>Assignment for a Business topic</u>: Research and prepare a paper and a presentation on businesses and advanced technology. What is hot, and what is not? Cite at least 3 reputable sources. Compare and contrast to the depictions in this book.

d) <u>Assignment to do a Startup:</u> Have the students prepare a paper about how they might startup a business in this realm. They must submit a sound Business Plan for the startup. They could also be asked to present their Business Plan and so should also have a presentation deck to coincide with it.

You can certainly adjust the aforementioned assignments to fit to your particular needs and the class structure. You'll notice that I ask for 3 reputable cited sources for the paper writing based assignments. I usually steer students toward "reputable" publications, since otherwise they will cite some oddball source that has no credentials other than that they happened to write something and post it onto the Internet. You can define "reputable" in whatever way you prefer, for example some faculty think Wikipedia is not reputable while others believe it is reputable and allow students to cite it.

The reason that I usually ask for at least 3 citations is that if the student only does one or two citations they usually settle on whatever they happened to find the fastest. By requiring three citations, it usually seems to force them to look around, explore, and end-up probably finding five or more, and then whittling it down to 3 that they will actually use.

I have not specified the length of their papers, and leave that to you to tell the students what you prefer. For each of those assignments, you could end-up with a short one to two pager, or you could do a dissertation length paper. Base the length on whatever best fits for your class, and the credit amount of the assignment within the context of the other grading metrics you'll be using for the class.

I mention in the assignments that they are to do a paper and prepare a presentation. I usually try to get students to present their work. This is a good practice for what they will do in the business world. Most of the time, they will be required to prepare an analysis and present it. If you don't have the class time or inclination to have the students present, then you can of course cut out the aspect of them putting together a presentation.

If you want to point students toward highly ranked journals in AI, here's a list of the top journals as reported by *various citation counts sources* (this list changes year to year):

- o   Communications of the ACM
- o   Artificial Intelligence
- o   Cognitive Science
- o   IEEE Transactions on Pattern Analysis and Machine Intelligence
- o   Foundations and Trends in Machine Learning
- o   Journal of Memory and Language
- o   Cognitive Psychology
- o   Neural Networks
- o   IEEE Transactions on Neural Networks and Learning Systems
- o   IEEE Intelligent Systems
- o   Knowledge-based Systems

GUIDE TO USING THE CHAPTERS

For each of the chapters, I provide next some various ways to use the chapter material. You can assign the tasks as individual homework assignments, or the tasks can be used with team projects for the class. You can easily layout a series of assignments, such as indicating that the students are to do item "a" below for say Chapter 1, then "b" for the next chapter of the book, and so on.

a) What is the main point of the chapter and describe in your own words the significance of the topic,

b) Identify at least two aspects in the chapter that you agree with, and support your concurrence by providing at least one other outside researched item as support; make sure to explain your basis for disagreeing with the aspects,

c) Identify at least two aspects in the chapter that you disagree with, and support your disagreement by providing at least one other outside researched item as support; make sure to explain your basis for disagreeing with the aspects,

d) Find an aspect that was not covered in the chapter, doing so by conducting outside research, and then explain how that aspect ties into the chapter and what significance it brings to the topic,

e) Interview a specialist in industry about the topic of the chapter, collect from them their thoughts and opinions, and readdress the chapter by citing your source and how they compared and contrasted to the material,

f) Interview a relevant academic professor or researcher in a college or university about the topic of the chapter, collect from them their thoughts and opinions, and readdress the chapter by citing your source and how they compared and contrasted to the material,

g) Try to update a chapter by finding out the latest on the topic, and ascertain whether the issue or topic has now been solved or whether it is still being addressed, explain what you come up with.

The above are all ways in which you can get the students of your class involved in considering the material of a given chapter. You could mix things up by having one of those above assignments per each week, covering the chapters over the course of the semester or quarter.

As a reminder, here are the chapters of the book and you can select whichever chapters you find most valued for your particular class:

Companion Book By This Author

Advances in AI and Autonomous Vehicles: Cybernetic Self-Driving Cars

Practical Advances in Artificial Intelligence (AI) and Machine Learning

by

Dr. Lance B. Eliot, MBA, PhD

This title is available via Amazon and other book sellers

Companion Book By This Author

Self-Driving Cars: "The Mother of All AI Projects"

by Dr. Lance B. Eliot, MBA, PhD

This title is available via Amazon and other book sellers

Companion Book By This Author
Innovation and Thought Leadership on Self-Driving Driverless Cars
by Dr. Lance B. Eliot, MBA, PhD

This title is available via Amazon and other book sellers

Companion Book By This Author

New Advances in AI Autonomous Driverless Cars Self-Driving Cars

by Dr. Lance B. Eliot, MBA, PhD

Chapter Title

This title is available via Amazon and other book sellers

Companion Book By This Author

Introduction to
Driverless Self-Driving Cars

by Dr. Lance B. Eliot, MBA, PhD

This title is available via Amazon and other book sellers

Companion Book By This Author
Autonomous Vehicle Driverless Self-Driving Cars and Artificial Intelligence
by Dr. Lance B. Eliot, MBA, PhD

This title is available via Amazon and other book sellers

Companion Book By This Author

Transformative Artificial Intelligence Driverless Self-Driving Cars

by Dr. Lance B. Eliot, MBA, PhD

Chapter Title

This title is available via Amazon and other book sellers

Companion Book By This Author

Disruptive Artificial Intelligence and Driverless Self-Driving Cars

by Dr. Lance B. Eliot, MBA, PhD

Chapter Title

This title is available via Amazon and other book sellers

Companion Book By This Author

State-of-the-Art
AI Driverless Self-Driving Cars

by Dr. Lance B. Eliot, MBA, PhD

Chapter Title

This title is available via Amazon and other book sellers

Companion Book By This Author

Top Trends in
AI Self-Driving Cars

by Dr. Lance B. Eliot, MBA, PhD

This title is available via Amazon and other book sellers

Companion Book By This Author

AI Innovations
and Self-Driving Cars

by Dr. Lance B. Eliot, MBA, PhD

This title is available via Amazon and other book sellers

Companion Book By This Author

Crucial Advances for AI Self-Driving Cars

by Dr. Lance B. Eliot, MBA, PhD

This title is available via Amazon and other book sellers

Companion Book By This Author

Sociotechnical Insights and AI Driverless Cars

by Dr. Lance B. Eliot, MBA, PhD

Chapter Title

This title is available via Amazon and other book sellers

Companion Book By This Author

Pioneering Advances for AI Driverless Cars

by Dr. Lance B. Eliot, MBA, PhD

Chapter Title

This title is available via Amazon and other book sellers

This title is available via Amazon and other book sellers

Companion Book By This Author

The Cutting Edge of
AI Autonomous Cars

by Dr. Lance B. Eliot, MBA, PhD

This title is available via Amazon and other book sellers

Companion Book By This Author

The Next Wave of
AI Self-Driving Cars

by Dr. Lance B. Eliot, MBA, PhD

This title is available via Amazon and other book sellers

ABOUT THE AUTHOR

Dr. Lance B. Eliot, MBA, PhD is the CEO of Techbruim, Inc. and Executive Director of the Cybernetic AI Self-Driving Car Institute, and has over twenty years of industry experience including serving as a corporate officer in a billion dollar firm and was a partner in a major executive services firm. He is also a serial entrepreneur having founded, ran, and sold several high-tech related businesses. He previously hosted the popular radio show *Technotrends* that was also available on American Airlines flights via their in-flight audio program. Author or co-author of a dozen books and over 400 articles, he has made appearances on CNN, and has been a frequent speaker at industry conferences.

A former professor at the University of Southern California (USC), he founded and led an innovative research lab on Artificial Intelligence in Business. Known as the "AI Insider" his writings on AI advances and trends has been widely read and cited. He also previously served on the faculty of the University of California Los Angeles (UCLA), and was a visiting professor at other major universities. He was elected to the International Board of the Society for Information Management (SIM), a prestigious association of over 3,000 high-tech executives worldwide.

He has performed extensive community service, including serving as Senior Science Adviser to the Vice Chair of the Congressional Committee on Science & Technology. He has served on the Board of the OC Science & Engineering Fair (OCSEF), where he is also has been a Grand Sweepstakes judge, and likewise served as a judge for the Intel International SEF (ISEF). He served as the Vice Chair of the Association for Computing Machinery (ACM) Chapter, a prestigious association of computer scientists. Dr. Eliot has been a shark tank judge for the USC Mark Stevens Center for Innovation on start-up pitch competitions, and served as a mentor for several incubators and accelerators in Silicon Valley and Silicon Beach. He served on several Boards and Committees at USC, including having served on the Marshall Alumni Association (MAA) Board in Southern California.

Dr. Eliot holds a PhD from USC, MBA, and Bachelor's in Computer Science, and earned the CDP, CCP, CSP, CDE, and CISA certifications. Born and raised in Southern California, and having traveled and lived internationally, he enjoys scuba diving, surfing, and sailing.

ADDENDUM

The Next Wave of AI Self-Driving Cars

Practical Advances in Artificial Intelligence (AI) and Machine Learning

By

Dr. Lance B. Eliot, MBA, PhD

———

For supplemental materials of this book, visit:

www.ai-selfdriving-cars.guru

For special orders of this book, contact:

LBE Press Publishing

Email: LBE.Press.Publishing@gmail.com

www.ingramcontent.com/pod-product-compliance
Lightning Source LLC
Chambersburg PA
CBHW021553210326
41599CB00010B/423